FACILITIES MAINTENANCE MANAGEMENT

Gregory H. Magee, PE

FACILITIES MAINTENANCE MANAGEMENT

Gregory H. Magee, PE

Copyright 1988

R.S. MEANS COMPANY, INC.
CONSTRUCTION CONSULTANTS & PUBLISHERS
100 CONSTRUCTION PLAZA
P.O. BOX 800
KINGSTON, MA 02364-0800
(617) 585-7880
6 7 8 9 0

ISBN 0-87629-100-0
Library of Congress Catalog Card Number 88-151879

TABLE OF CONTENTS

FOREWORD

Every facility, old or new, requires maintenance. Interior and exterior wallcoverings, operating systems, machinery, flooring, and any other building part all require some degree of maintenance to ensure the longest possible life for each component. Controlling maintenance includes managing costs as well as productivity monitoring and work tracking. However, the maintenance of these components must be fully understood in order to be controlled. *Facilities Maintenance Management* provides forms, formulas, and methodologies to manage these aspects of the maintenance effort.

Chapter 1, "What is Facilities Maintenance Management?", establishes reasons for maintenance management. Methods are introduced for determining which maintenance activities are necessary, establishing maintenance objectives, and categorizing maintenance activities. The chapter ends with some typical universal maintenance goals and specific objectives.

Chapter 2, "Organizing and Staffing for Maintenance", details necessary staffing requirements for small, intermediate, and large organizations. This discussion includes every position from the maintenance manager to the maintenance mechanic.

Chapter 3, "Engineering Considerations", details the role of engineering in facilities maintenance. This begins with the initial design — designing for maintainability — and includes methods for selecting an architect or engineer and managing new construction, modifications and improvements. The chapter ends with a discussion of troubleshooting.

Chapter 4, "Cost Estimating and Budgeting", details the elements of maintenance costs: labor, materials, parts, supplies, tools, and equipment. Unit price estimating techniques are also explained.

Chapter 5, "Cost-Based Maintenance Decisions", presents two methods for the economic comparison of alternatives: the *present value method* and the *uniform annual cost* method, with formulas and examples.

Chapter 6, "Identifying the Maintenance Workload", Chapter 7, "Evaluating and Executing Maintenance Work", and Chapter 8, "Controlling the Maintenance Effort", discuss further ways in which to manage actual maintenance workforce and workload.

Chapter 9, "Computerized Maintenance Management", discusses options for purchasing computer systems. A discussion of implementation follows.

Chapter 10, "Contracting for Maintenance Services", discusses when and how to contract for services. Types of contracts and services typically procured are outlined.

Facilities Maintenance Management ends with Chapter 11, "Preventive Maintenance", describing the rationale and methodologies for implementing this most important aspect of facilities maintenance.

Facilities Maintenance Management will be useful to anyone responsible for overseeing the maintenance effort of a facility, large or small, from the facility owner to the maintenance manager.

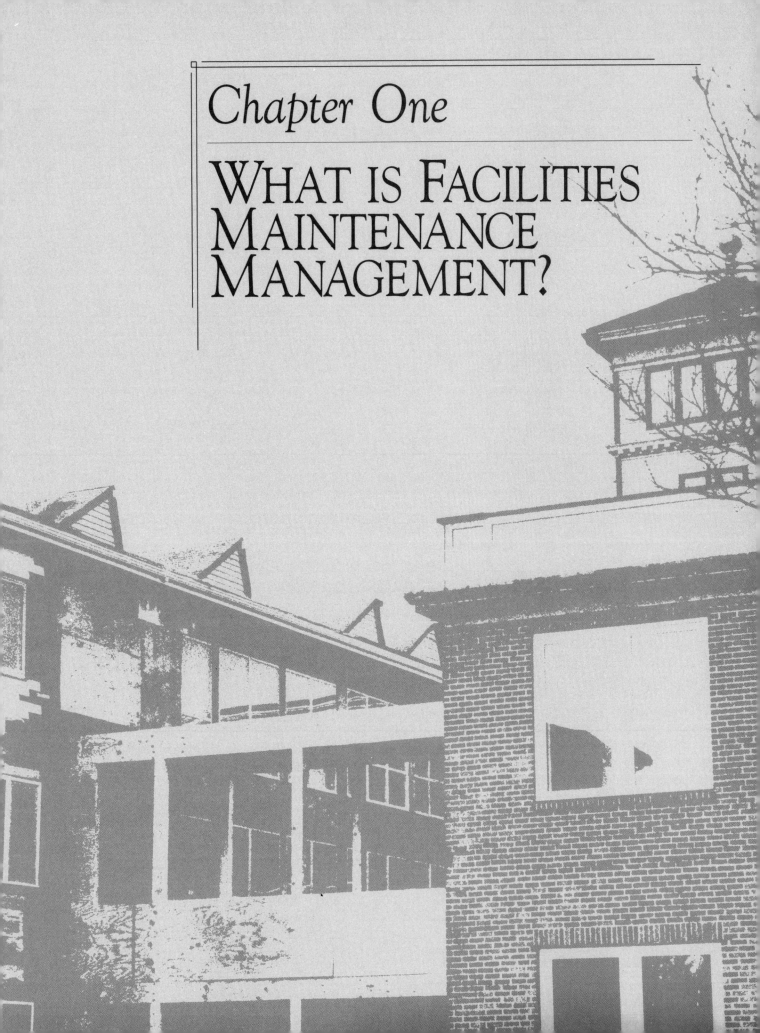

Chapter One

WHAT IS FACILITIES MAINTENANCE MANAGEMENT?

Chapter One

WHAT IS FACILITIES MAINTENANCE MANAGEMENT?

Facilities maintenance management is not a new or difficult concept to understand. By the age of four, most persons have had and comprehended the rudimentary experience of taking care of some possession, suffering when it functioned improperly, struggling to make it work, enjoying that possession when nothing was amiss, and paying a price when the possession was abused. Like a child with his favorite toy, the facility manager has the same fate and, all too typically, the same level of success and failure.

The difficulties inherent in the proper maintenance of a facility, whether it is a hospital, elementary school, industrial plant, office building or shopping mall, are compounded by the fact that it takes no specific level of education to identify when maintenance is working or when it has failed. To the facility manager, the world is full of critics, thanks are infrequent, perfection expected. The link between that perfection and profitability is tenuous, while the link between poor performance and losses is immediate. If the facility manager ever doubts that his or her job is important, he or she should try turning off the air conditioning for about an hour on a hot summer day.

The proper maintenance of a facility is easy to define and easy to evaluate. The importance of maintenance is easily established. Yet achieving an acceptable level of maintenance is extremely difficult. Generally, if everything is working properly, if cleanliness is up to standards, if the work force and clients are all happy, then one can expect that the facility manager is over budget. The facility manager therefore is a juggler; weighing the options of maintenance, repair, and replacement against their respective costs; setting and resetting priorities; "putting out fires"; and, on a good day, feeling tremendous pride of accomplishment.

Life Cycle Stages of a Facility

The life of a facility, whether of one building or of an extensive complex, can be divided into several finite stages: design, construction, occupancy and use, repair, rehabilitation, and disposal. Similar stages apply for individual systems or components within a facility.

3

Within each stage there are additional definitive sets of actions required of the facility manager or owner and each of these actions has an attendant cost. For example, since all facilities or their component parts arise from some recognized need, it can be assumed that the facility (or component) is justified and must remain in place and in operation. Whether the need comes from an increase in population, initiating the construction of a new school, or is in response to a corporate objective to expand profits, causing construction of new office or plant facilities, the objective of providing that operating facility on schedule at the lowest possible cost is always paramount.

Prior to the birth of a new facility, all estimated potential costs are controllable. As soon as the life of the building or component begins, however, the control (or flexibility) of the total *life cycle cost* of the facility diminishes. Once the design for the facility is fixed, so too are the majority of the anticipated operating and maintenance costs. Few opportunities will arise in the life of the facility to markedly change these operating, or life cycle, costs. Therefore, it is imperative to understand these maintenance and operating costs and to address them during the design stages. When a facility already exists, it is imperative that the maintenance effort be studied to uncover any possible reductions in operating and maintenance costs. These cost savings can compound for many years.

Once a facility or improvement to a facility is approved, the first step is to firmly define the need. With the specific needs in mind, the new facility is designed and construction begins. At the completion of construction, the facility is occupied and used. From this point forward the facility must be maintained in a usable state to assure that the original need is continually satisfied. When this need is no longer present, or the function obsolete, the facility or component is disposed of. Otherwise it will, at some point, require refurbishing or remodeling.

During the life of a facility the stages of design, construction, and disposal are brief periods when compared with the operation and maintenance stage. Initial construction, however, as the most costly stage, receives a "lion's share" of attention from the organization. Operation and maintenance is often accepted as a given, but cursed for its nagging presence. There is no glamour in maintenance, but, as a major long term cost, maintenance must be controlled. To be controlled, maintenance must be understood.

Establishing Maintenance Activities

Facility maintenance is the set of ordered activities which, when properly managed, allow for the continual operation of a facility. These activities include decisions and actions made by members of the maintenance team. The maintenance team, however, must first be properly staffed and trained. Staffing will be further discussed in Chapter 2.

The maintenance effort begins with organizational decisions concerning the desired level of maintenance. Policy regarding the *source* to be notified for completion of the maintenance work must then be established. Next, the actual extent of the maintenance work to be performed must be identified, work crews scheduled, and the actual work completed. Anticipated costs are estimated and a maintenance budget established.

In the background of these readily identifiable, or *direct*, activities are numerous *indirect* functions which proceed simultaneously to ensure that the proper level of maintenance is sustained. These background activities include spare parts inventory and ordering, engineering support, accounting, payroll, and payment and billing for services received or rendered. Perhaps the most important indirect maintenance activity is the continuous comparison of budgeted maintenance costs to actual expenditures—cost accounting and control.

Establishing Maintenance Objectives

The facility manager may have a difficult time keeping his maintenance program on track if there is no end goal or set of guideline objectives. There are no firm or "boilerplate" standards which can be applied to all facilities. Maintenance goals vary considerably, depending on the intended use of the facility and the business of the user. Some typical, broad maintenance goals are listed below:

- Provide a safe, clean, healthy environment conducive to providing educational programs to elementary school students.
- Provide a reliably safe and clean environment for providing inpatient and out patient health care services.
- Provide an excellent level of personal comfort to guests staying at the hotel.
- Provide an adequate working environment for various tenants in an office building.
- Provide the minimally acceptable level of work necessary to maintain the waterproof integrity of the warehouse.

The simple goal statement typically defines the emphasis or direction for the maintenance effort and gives some indication of the intensity of effort required. Simple goals must be backed by more specific objectives. These objectives should address the various components of the entire maintenance program.

Paramount in developing each stated objective is the cost of implementing each objective. Maintenance activities must be performed at the proper frequency and at the least cost. A set of maintenance objectives for any particular facility, therefore, should address the costs of each of the major components of facility maintenance. The following is a brief description of the various components of a facility maintenance program and some generalized objectives for each. These objectives can easily be modified to accommodate any individual facility maintenance program.

Direct Maintenance Work

Direct maintenance work includes activities which preserve or restore the function of the facility. This category is divided into the following subdivisions: housekeeping, general maintenance, preventive maintenance, repair, replacement, improvement, modification and utilities. Every direct maintenance work item can be placed in one of these categories. This breakdown also defines, to a certain extent, the type of resources required to perform the work. The names of these categories are likely to vary slightly from one organization to another; but the basic elements, as described, must be present in every facility maintenance program.

Housekeeping

Housekeeping is that group of activities which make the facility presentable and fully usable to its clients, preserving the proper day to day operation of a properly functioning facility. Included are basic cleaning of spaces, emptying of trash receptacles, replacement of towels and toilet paper, sweeping, mopping, and dusting. Housekeeping tasks are limited to those performed by unskilled labor on a frequent basis, generally daily.

A sample objective for the housekeeping function might be:

> Ensure that all spaces are cleaned once within each 24-hour period.

The objective might go so far as to define a limit on the amount of time a piece of discarded trash would be allowed to remain on the floor. Such a severe objective implies continual inspection and a large, dedicated work force. Yet, for some facilities it is apparent that such an objective is in place.

General Maintenance

General maintenance might also be described as *infrequent housekeeping*. A general maintenance activity often requires somewhat more skill than housekeeping, and often uses specialized equipment. Typical general maintenance activities include stripping and re-waxing floors, repainting walls and and trim, spring planting flower beds, sweeping roads and parking lots, or steam cleaning carpets. General maintenance improves or preserves the appearance of a facility and is accomplished at discrete intervals based on seasonal considerations, accumulated experience, or aesthetic preferences. However, many general maintenance activities, if continually neglected, may lead to premature failure of the facility component.

Also included within the category of general maintenance are *nuisance work items* such as replacing faucet washers, tightening loose door or window hardware, tightening valve gland packings, adjusting door closers, replacing light bulbs, and lubricating hinges.

Along with housekeeping, general maintenance is often the most visible of the various maintenance activities. The need for general maintenance is most frequently identified and reported by the facility user. General maintenance activities are rarely critical to operation of a facility. If ignored, however, more severe problems usually arise. Further, the users may cease reporting problems if no prompt action is ever taken to preserve this vital stream of user input. A suitable objective for general maintenance might read:

> Promptly respond and repair minor discrepancies in facility functions.

Preventive Maintenance

Preventive maintenance is any work performed to an operational device or facility to continue operating at its proper efficiency without interruption. Preventive maintenance activities are performed at regular intervals, usually by a skilled work force. As an individual category, preventive maintenance is significantly different from general maintenance. The interval between preventive maintenance actions on a particular component is established by manufacturer's recommendations, empirical measurements of degrading performance, or the impending failure of an unmaintained piece of equipment. When preventive maintenance is continually neglected, dramatic and costly failures often occur. For this reason, a formal preventive maintenance program should be a high priority. In the interest of cost, however, care must be taken not to allow a state of "over maintenance" to creep into the program. Once the preventive maintenance program is formally in place, it must be continually examined to determine if the cost of frequent maintenance exceeds the cost of downtime and repair if no maintenance were performed. A suitable objective might be:

> Establish a program of routine inspection and services of equipment to prevent premature failures.

Repair

Repair work involves restoring to operation some component of the facility after it has failed. It is the "headline maker" of facilities maintenance, as failures rarely happen at convenient times. Most often the repairs must be made immediately, at the expense of other scheduled maintenance.

In establishing objectives for completing repairs it is often necessary to set priorities based on the urgency of need for the repair. These priorities establish the desired maximum time required to complete repairs. Thereafter, as failures occur, the repair is classified by priority and work is scheduled accordingly. Work which is not immediately required (low priority) is often placed on a backlog for future scheduling and accomplishment. A typical objective for repair work might be to:

> Complete ninety percent of all repair work within prescribed limits.

Although the various priority groupings have finite desired time limits for accomplishment, the actual assignment of priorities to work is subjective. Therefore it is further necessary to establish criteria for each priority grouping. For example, this may seem redundant, assigning a priority to a project rather than setting a simple time limit for completion, but it is necessary to direct the scheduling of work by shop personnel. Cost comparisons are often made at this stage when the magnitude of the failure indicates that replacement should be considered.

Replacement

Replacement, as a distinct work element, is confined to a program of planned replacement of facility components. It may be further limited to major components such as air conditioning compressors, furnaces or hot water heaters. Replacement is performed when the equipment has reached the end of its useful life; when it no longer can perform due to degradation of its internal components and repair is no longer cost effective. Included under the title of replacement would be the major rebuilding of any component, since rebuilding also restores performance.

Ideally, in a maintenance program with sufficient historical data on similar components and a means for noting component degeneration, the replacement of any facility component would be scheduled to occur just prior to total failure. Replacement is, therefore, the final step in an orderly maintenance program which has extracted the most cost effective use out of a component.

Although the decision to replace a piece of equipment is generally inevitable, it is not without a wide variety of options. Accordingly, a program of planned replacement should revolve around the costs of the equipment, its installation, and its maintenance. The replacement program presents a unique opportunity to the facilities maintenance manager. Since component degeneration mandates some form of replacement, the manager can analyze the impact of using a different component which might result in a lower life cycle cost. In other words, an item that is initially more expensive, might require less repair and less frequent replacement.

The replacement objective for any facility might read:

> Execute a program of planned replacement of major facility components which replaces failing equipment (before failure) with new or rebuilt components which have a lower life cycle cost.

Improvement

Improvement projects enhance the proper operation or reduce the operating costs of a facility. These projects may include the installation of energy and utility conserving devices such as flow restricting faucets, thermal insulation, or more efficient heating or cooling system components. Replacement of properly operating but maintenance intensive equipment with similar but more reliable products is included in this category.

Life cycle cost is the essence of any improvement project. However, while any project which provides for reduced long term costs is worthwhile, improvement projects are generally costly and maintenance budgets may not be sufficient to sustain them.

While other maintenance activities are dictated by the facility or component in place, improvement projects are often initiated by maintenance management personnel. It is the responsibility of the maintenance staff to continually seek methods to reduce operation costs. A maintenance manager should seek to identify cost saving projects for consideration by upper management. A typical objective in this area might be:

> Identify and execute any improvement project which will provide a payback of the initial investment in three years or less.

Modification

Modification projects alter the basic facility or facility component to accommodate a new function or corporate initiative. Modification projects differ from improvement projects by their point of origin. Modification projects are initiated from outside of the maintenance staff whereas improvement projects are initiated by maintenance staff. The estimated cost of a modification or alteration must be considered well in advance, since corporate decisions concerning new initiatives are often based on cost. Since the decision to go ahead with a modification project relies heavily on the estimated costs, a suitable objective might be:

> Accurately estimate the costs of modification projects and to complete such projects at or under budget.

Utilities

The direct work elements previously discussed have generally required the expenditure of on-site labor. Utilities are included as a direct work item since many facilities generate their own utilities. The utilities work element includes furnishing electrical power, potable water, centrally produced heat or cooling fluids, collection and disposal of sewage and other wastes, and collection and disposal of storm water.

Utilities are usually provided by local municipal utility systems for small facilities. In this case, the involvement of a facility or maintenance manager is minimal. For a small facility, the utility billings are simply verified and certified and the utility company called when service is interrupted and/or repairs are required.

In larger facilities, this may involve the full time operation of potable water wells and treatment plants, full electrical generating equipment, massive central boilers for distributed steam for heat or processing, and in many cases, the operation of a sewage treatment and disposal facility. Such systems become mini-facilities in themselves, requiring "round the clock" attention, and contribute substantially to the overall workload for maintenance and repair of the facility as a whole. For this reason, internally operated utility systems often represent a significant portion of the operating costs of a facility. Of course, the decision to internally provide these utilities was originally made because it was either cheaper, more reliable, or the service in question was not available in sufficient quantity from the municipal source.

Indirect Work Elements

Indirect work elements are the activities which facilitate the direct work previously outlined. While not performed directly on a facility, these elements should be present in any cost effective program. Indirect work elements include:

- work identification,
- cost estimating,
- purchasing,
- supplies and inventory control,
- scheduling,
- work tracing and monitoring,
- facility and equipment histories, and
- engineering.

Work Identification

All physical work performed to a facility has a point of origin. Preventive maintenance, for example, is a planned activity whose frequency is either firmly established to prevent failure or is scheduled seasonally, such as switching over from heat to air conditioning. Housekeeping is a regularly scheduled activity. General maintenance and repair work, however, are not scheduled and must first be noticed and then reported to the maintenance staff. In the course of preventive maintenance and formal daily rounds of equipment, any irregularities noted by the staff should be either corrected or logged for future repair. Costs prohibit continual maintenance staff inspection of all facilities and spaces, thus the facility user must be relied upon to identify and report repair and general maintenance discrepancies.

A non-functioning item is often first noted by the user. Yet, if the user is to be relied on as a source of identifying needed repairs, it is imperative that all users have an easy mechanism for reporting broken items. Without a simple method for contacting the maintenance staff, the user soon forgets to report the casually noticed items, assuming that noticing and reporting is someone else's responsibility. Since minor problems, when neglected, never cure themselves and invariably grow into major problems, it is essential that a readily accessible method for reporting problems be an integral part of a total facility maintenance program. A reporting system is most typically established by use of a work receptionist who receives telephone notice of any discrepancy.

Major work performed by the maintenance staff under the categories of replacement, modification, or improvement is identified through various means. Replacement is predicted by the *rate of decay* in performance of the facility or equipment. Modifications are prescribed by the user to accommodate new or different facility functions. Improvements are the only projects which are voluntarily initiated. Although improvement projects are often facilitated by a simultaneous need for major replacements, the primary reasons for executing improvement projects are long term monetary savings or increased profits.

A formal inspection program is another means for identifying work to be performed. This program might entail hourly checks of critical equipment or annual checks of unoccupied, inaccessible spaces in a building. The frequency of inspections would depend on the impact of failure of the inspected equipment or facility, should the early stages of degeneration go undetected.

A formal maintenance objective for proper identification of work to be performed might read:

> Conduct a program of formal inspection and facility monitoring scheduled to identify critically needed maintenance work, coupled with an open communication channel for non-staff members to report facility discrepancies.

Cost Estimating

Cost estimating is an integral part of all facets of facility maintenance. The role of cost estimating begins with the development of the annual operating and maintenance budget; it is necessary to predict the frequency and scope of the maintenance work to be performed during the year. Historical records and trends are the most valuable tools available when preparing such a budget. From historical data the overall cost, or budget, for each area of work is developed.

As individual repair work items are identified during the course of the year, it is necessary to produce a more refined estimate of the costs (to measure the progress of each individual task). A cost estimate may determine whether an item is repaired or replaced. Cost estimates play a similar role when evaluating the feasibility or desirability of modifications or improvements to the facility.

Accurate costs depend on several variables. The prevailing wage rate, expected level of productivity, material costs, equipment costs, and overhead costs must be given full consideration. The relative predictability of these cost elements vary. Wage rates, for example, are usually very stable, while productivity levels are often unknown and highly dependent upon individual circumstances and work conditions. Thus, if the actual cause of the component failure is not immediately evident, the repair costs may be extremely difficult to estimate before the particular repair is completed.

Since cost estimating is extremely difficult and often highly dependent on historical cost data and past experience, the task of preparing estimates is generally assigned to a limited number of persons within an organization. A management objective regarding cost estimating might be:

> Develop a formal method for estimating the costs of all maintenance work to allow for both long and short term control of the maintenance budget.

Purchasing, Supplies, and Inventory Control

As maintenance work is carried out, there is a constant requirement for materials and supplies to sustain the work effort. This necessitates a formal system for predicting the materials required, procuring those materials, and maintaining accountability for those materials. The total material requirements and proper inventory levels for a budget period depends on two factors: the frequency of need for the materials or parts in question, and the impact of not having a sufficient inventory of parts on hand on facility operations. Routine preventive maintenance tasks require a predictable type and quantity of materials. Repairs, being unpredictable, may cause interruptions to facility operations if appropriate parts are not readily available.

In addition to the procurement of materials and parts, it may be necessary to purchase services from outside repair specialists. Formal purchasing practices must be established to ensure that contractual obligations are established and met by servicing personnel. Where purchasing flexibility exists for selecting materials, suppliers, and service contractors from a source other than the lowest bidder, the maintenance manager must be prepared to assess the value of the product versus the price paid. The adage "you get what you pay for" is often very true in facility maintenance. For this reason, life cycle costs should be considered at each purchase or contract. A suitable objective for purchasing, sales, and inventory control might be:

> Establish and follow a policy of purchasing and inventory level which ensures that proper materials, parts, and services are available to maintain full facility operation.

Cost Accounting and Control

Although most companies or organizations have formal accounting systems which concentrate on payables and receivables, it is necessary for the maintenance manager to have a parallel system to measure the costs of the various operations and transactions on a more frequent and more definitive basis. As computers gain a hold in an organization it is frequently possible to merge the two requirements. Lacking a suitable composite system, however, the manager may find it necessary to develop a cost management system to track costs by individual job, discrete facility functional area, direct work type, assigned work force or service contractor, and individual facility, equipment, or component. Additionally, the scope of the total maintenance effort may dictate subdivision of cost accounting to individual departments within the maintenance staff.

The purpose in cost accounting is to accurately measure the ongoing and historical costs of each maintenance activity. From this measurement, the maintenance manager can make decisions to redirect or reallocate resources to mitigate increasing costs or lessen stable costs. For example, excessive overtime costs might lead to hiring or reassigning additional full-time employees. Therefore, timeliness of reporting ongoing costs is an essential factor. The desired frequency of measuring and assessing ongoing costs varies depending on the risks associated with not measuring the costs. For example, preventive maintenance activities (see Chapter 11) can be accurately estimated, are unlikely to vary significantly from the estimate, and need not be monitored frequently. Repair project costs, however, may fluctuate hourly and require frequent monitoring. The potential for costs to stray from the budgeted amount is limited for fixed price service contracts but can be great for undefined "open, inspect, and repair" work. Since the maintenance manager's time is *always* limited, efforts should be concentrated on the highest risks. Since the possibility for long term over-budget problems can develop by a compilation of slight overages, the manager must also have a method for tracking cumulative costs within the various categories mentioned above. A suitable objective for cost accounting and control might be:

> Maintain a system for tracking the ongoing and cumulative costs of the maintenance effort, which supplies timely information for budget comparison and management decision-making.

Scheduling

Since maintenance work elements are both planned (preventive maintenance, improvement, modifications, housekeeping) and unplanned (repairs), the maintenance manager must be an efficient scheduler. It is necessary to provide sufficient scheduled workers for the jobs planned while maintaining sufficient flexibility to handle most unforeseen events. To adequately plan for anticipated and unanticipated work items, two elements must be scheduled: the work and the workers. As noted earlier, preventive maintenance work is scheduled at defined intervals based upon need and risk. These work items must, therefore, be set down as fixed work and staff provided for their accomplishment. Similarly, known or surprise repair work must be covered. Once identified, individual workers must be assigned specific tasks for finite periods. This scheduling is based on the accurate productivity estimates established for each of the tasks.

From the known workload of planned tasks and the historically predicted incidence of repair work, an optimum staffing level (based on historical data) is established. Ideally, all planned work is accomplished and the backlog of repair work does not grow, if the staffing is adequate. Generally it is better to be slightly understaffed, resorting to overtime, than to be overstaffed which may lead to decreases in productivity as workers slow down to fill the day. A suggested scheduling objective might be:

> Schedule and assign workers to specific tasks to gain maximum utility from the wages paid while sustaining the necessary amount of work.

Work Tracking and Monitoring

In addition to monitoring the costs of the maintenance effort, attention must also be paid to ensuring that all work is completed in a timely manner. Since deferred preventive maintenance or neglected repair work can frequently lead to costly repairs or facility downtime, a system for tracking the identification and accomplishment of all work is imperative. If client-identified minor repairs are not accomplished in a reasonable amount of time, the client or user may soon decide it is useless to report minor problems. If minor problems are not reported, they could lead to major problems, therefore, the work tracking system must be timely and complete. The method for tracking work should establish defined intervals for workers or foremen to report completion. The length of the interval depends on the anticipated effect of *not* completing the work. For example, a weekly report of all routine work may be sufficient for certain repairs, while hourly reports for emergency repairs may be too far apart. A primary reason for having a work tracking system is simply to ensure that no required work is forgotten.

The work tracking system should be logically tied to both the cost accounting and work scheduling systems. It must have the capability to track ongoing work progress (and costs) which exceed regular reporting intervals. The system should also be able to provide feedback to both the facilities maintenance staff and the facility users concerning work status.

The objective governing work tracking might be:

> Utilize a formal method for keeping track of the status of costs and progress for all planned and unplanned maintenance and repair work.

Facility and Equipment Histories

In order to properly predict and adjust the maintenance program for a facility, the manager must have an intimate knowledge of the life of the particular facility and, therefore, of facility components. Only with a total building history can proper decisions be made regarding the best course of maintenance actions for a given problem. Accurate recordings of actual construction, all modifications, and improvements should be kept. Major and recurrent minor repairs must be tracked. Periodic inspections must be made and the current state of the facility assessed and recorded. The same types of records should be kept for major equipment components. The purpose of accurate records of the total facility is to enable immediate troubleshooting and repairs.

A proper equipment history should record the make and model of the equipment, date of installation, all major repairs, all preventive maintenance, routine parts replacements, and any continuing engineering measurements of equipment performance. Additionally, where the work force is stable, the name of the worker performing the repair or preventive maintenance should be noted. The equipment history is then used to predict the need for eventual replacement or rebuilding of the component, or when trying to diagnose an unexpected failure which can often be traced to a recent repair.

A reasonable objective for facility and equipment histories might be:

> Maintain current and accurate records of the present status of all facilities and equipment including initial construction or installation, repairs, and modifications to aid in timely repair and proper replacement of components.

Engineering

The previous direct maintenance work elements and associated indirect work can usually be accommodated without a formal in-house engineering staff. Many work elements, however, require engineering input at some stage. Engineering services include assessing facility and component performance, designing modifications to facilities or equipment, preparing engineering drawings and specifications for repairs, and troubleshooting major component or system failures.

Many large facilities may operate for long periods of time without formal engineering assistance. When, however, the facility manager needs technical assistance it is often expensive since the engineer must gain familiarity with the facility. For this reason, it is desirable to maintain an engineer or engineering firm on continual retainer, requiring that firm to stay conversant with the facility functions and physical components. The engineering disciplines predominately involved in facility maintenance are mechanical engineering, civil engineering, architectural engineering, and electrical engineering.

The need for engineering input is usually proportional to the size and complexity of the facility. If a facility were to grow in an orderly fashion, adding buildings and components, there might be a time at which the services of an engineering consultant become so frequent and the costs so great that in-house engineering capability may be more practical and efficient. In addition to pure cost, the need for more immediate analysis and response than that which outside engineering services can render may dictate the decision to expand the maintenance staff with more engineering talent. There is no set facility size that dictates in-house engineering capability; the need varies based on facility function as well. The governing objective for providing engineering services might be:

> Maintain sufficient personnel trained in the engineering disciplines to design, analyze, and troubleshoot maintenance and repair projects for the facility.

Universal Maintenance Goals and Objectives

The following is a generic set of goals and objectives for a maintenance organization. These statements can be revised to meet the specific needs of any facility, its intended use, and the strategic goals of the company.

Overall Maintenance Goal

Provide economical maintenance and housekeeping services to allow the facility to be used for its intended purpose.

Specific Maintenance Objectives

- Perform daily housekeeping and cleaning to maintain a properly presentable facility.
- Promptly respond and repair minor discrepancies in the facility.
- Develop and execute a system of regularly scheduled maintenance actions to prevent premature failure of the facility and its systems and components.
- Complete major repairs based upon lowest life cycle cost.
- Identify design and complete improvement projects to reduce and minimize total operating and maintenance costs.
- Operate the facility utilities in the most economical manner while providing necessary reliability.
- Provide for easy and complete reporting and identification of necessary repair and maintenance work.
- Perform accurate cost estimating to ensure lowest cost solutions to maintenance problems.
- Maintain a proper level of materials and spare parts to support timely repairs.
- Accurately track the costs of all maintenance work.
- Schedule all planned work in advance, and allocate and anticipate staff requirements to meet planned and unplanned events.
- Monitor the progress of all maintenance work.
- Maintain complete historical data concerning the facility in general and equipment and components in particular.
- Continually seek workable engineering solutions to maintenance problems.

This list provides general goals and objectives of value to any maintenance manager. In actual use, however, these objectives would be much more specific — tailored to a particular facility. For example, where a general description such as "timely" appears, the facility manager should be more descriptive, such as "daily", "weekly", etc. Once refined and detailed, the objectives should be given to all key members of the maintenance organization. From that point forward, the objectives should be generally followed. Within days of devising the specific objectives for a facility, it will be obvious that the objectives cannot substitute for common sense. Even very specific objectives can not fit each unique situation. But the alternative to setting down a statement of goals and objectives is to have the maintenance organization wander like a ship without a rudder, always making progress but perhaps not always in the proper direction.

Summary

In the life of a facility, its occupancy, use, and operation present many responsibilities for the facility manager. Proper maintenance for the facility means keeping the facility and its components in good condition, ready to serve their intended purpose. This involves both direct physical work on the facility and indirect supporting work by the maintenance staff. The total maintenance effort should be guided by an internally developed and formally defined set of goals and objectives. At all times, decisions regarding maintenance work should be based on life cycle cost implications.

Chapter Two

ORGANIZING AND STAFFING FOR MAINTENANCE

Chapter Two

ORGANIZING AND STAFFING FOR MAINTENANCE

In order to maintain any facility, a formal organizational structure must be defined. The scope of this structure depends on the size and complexity of the facility to be maintained, the skills required to maintain it, and by the desired supervisory span of control. Once the organizational structure is defined, the actual hiring of employees proceeds according to the type and level of skill required by each position. Once an organization is staffed, the evaluation, training, and continued development of the staff requires continual attention. Suggested methods of handling each of these tasks are presented in this chapter.

Organizing for Maintenance

A maintenance department is organized according to the direct and indirect maintenance activities of the organization (see Chapter 1 for definitions of these terms). The size of each subdivision of work (direct and indirect) depends on the required volume of each work type. In the smallest of facilities, a single maintenance employee may have responsibility for both direct and indirect maintenance activities. On the other end of the spectrum, a facility that encompasses millions of square feet of buildings and thousands of acres may have several work crews of a particular type of craftsman for only a small portion of one of the work divisions. The development of any maintenance organization, however, involves three components: overall facility management, maintenance management, and maintenance workers. Although every organization has all three, smaller organizations may have only one or two employees who cover all three components. The size and scope of a maintenance organization is dictated by the following factors:

- Size of the facilities to be maintained
- Complexity of the facility
- Costs associated with facility failures
- Defined maintenance objectives
- Management philosophy toward maintenance
- Relative "political" strength of the maintenance management staff within the total facility organization
- Age of the organization
- Current industry trends in maintenance staffing

Out of the above factors emerges a maintenance organization of a particular size and scope necessary to meet the facility's operational and maintenance objectives. There are no standard names or labels describing the differing sizes of maintenance organizations, but, for the purpose of describing the relationships between management, maintenance management, and maintenance execution, three basic classes of organizations are sufficient. These three classes are large, intermediate, and small maintenance organizations.

Large Maintenance Organizations

Large maintenance organizations often have numerous formal management and engineering departments and a large maintenance staff. The maintenance staff is usually part of a branch called *Facilities Engineering*, which is subdivided into shops by types of work performed. The facilities engineering branch consists of two to four subunits addressing engineering design, maintenance control, maintenance execution, and utilities management. Generally, a professional engineer heads the facilities engineering branch. An example organizational chart for a large maintenance organization is shown in Figure 2.1.

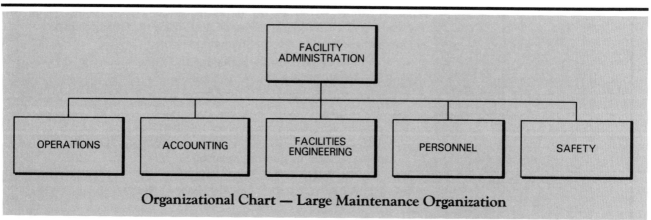

Organizational Chart — Large Maintenance Organization

Figure 2.1

20

Major facilities with many diverse and large buildings and substantial property require a large maintenance organization. Examples include a university campus, a major medical center, a metropolitan airport, a commercial office or industrial park, and a large manufacturing or processing plant.

Facilities Engineering

The Facilities Engineering Office is headed by a professional engineer, usually entitled the *facilities engineer*. Since the coordination between design and maintenance is a major factor in reducing and controlling maintenance costs, the facilities engineer ensures that the designers and maintenance employees work closely together. The design group produces new designs that are easily maintained and that fit within the current maintenance scheme and spare parts inventory. The maintenance work force provides feedback to the design staff about maintenance problems encountered with recent designs. The facilities engineer also makes final engineering decisions, and supervises both the professional design staff and the labor-intensive maintenance staff. The Facilities

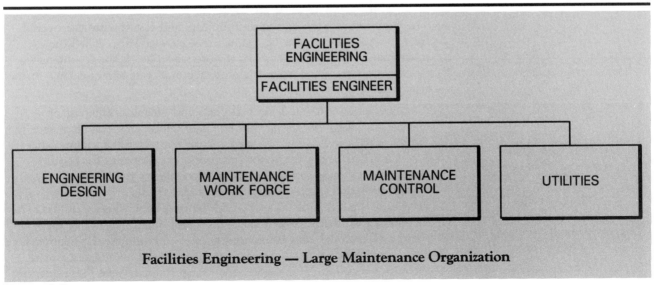

Facilities Engineering — Large Maintenance Organization

Figure 2.2

Engineering Office interfaces the engineering design and maintenance components with the overall operation and goals of the facility. Figure 2.2 illustrates the organization of a typical facilities engineering department.

Engineering Design Staff

The *Engineering Design Staff* consists of professional architects and engineers who produce designs for improvement and modification projects. The Engineering Design Staff is headed by a professional engineer or architect who reports directly to the facilities engineer. As discussed in later chapters, the decisions made by this staff directly affect the maintenance costs of the facility. This branch varies in size depending on the magnitude of the facility and the frequency of need for engineering designs for modification. In facilities where the need for new designs is infrequent and insufficient to support a full staff, engineering services are procured from outside architectural engineering firms or consultants. However, some engineering talent is usually required for all large facilities for troubleshooting and advising the maintenance foremen and workers. In such cases, the engineers are often a subgroup within the Maintenance Control Branch. Figure 2.3 is a basic organizational chart for the Engineering Design Branch of a large maintenance organization.

Utilities Branch

The *Utilities Branch* is present only in facilities that are engaged in the self-generation of major utility services. The utilities branch is usually responsible for the oversight of potable water treatment plants, electrical power plants, central steam or cooling plants, and waste water treatment plants, and is traditionally headed by an electrical, mechanical, or environmental engineer. The utilities branch basically controls the quality and quantity of the various plant outputs. Quality control is generally dictated by both facility needs and the governing local, state, and federal regulations. For example, the operation of potable water treatment and waste water treatment plants requires a state-licensed operator in most areas. The utilities branch is often responsible for the proper handling and disposal of hazardous waste. Providing "cradle to grave" tracking of hazardous waste and complying with water and sewage treatment standards requires a wide variety of report generation to the various local, state, and federal agencies involved.

Maintenance Control Branch

The *Maintenance Control Branch* monitors and coordinates the overall maintenance effort. A *facility maintenance manager*, also called the maintenance manager or chief of maintenance, heads the Maintenance Control Branch. The facility maintenance manager is frequently a former maintenance supervisor.

The Maintenance Control Branch plans and estimates the cost of maintenance activities. A *planner/estimator* within this branch is usually in charge of the planning and cost estimating for requested maintenance projects. The planner/estimator produces cost estimates for higher management. These estimates are used to evaluate the merit of modification and improvement projects and are also reviewed to determine the scope of the work and the type of workers required. Then the planner/estimator determines the correct quantities and types of personnel, material, and equipment needed to complete the maintenance task. This is done *before* the task is assigned to the maintenance work force for execution. The maintenance shop foreman uses these quantity

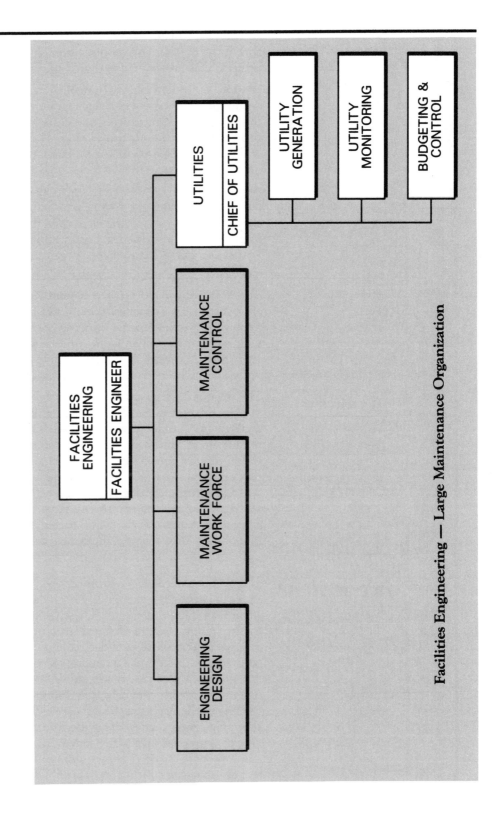

Figure 2.3

Facilities Engineering — Large Maintenance Organization

and labor estimates to schedule daily work assignments. A *scheduler* within the maintenance control branch usually coordinates and schedules projects that involve more than a single shop.

Also in the Maintenance Control Branch, a *work tracking and monitoring* group identifies new work through a trouble desk, receiving input from the facility users, producing worklists for the various shops, and tracking project assignments to completion. The work tracking and monitoring group is also responsible for the maintenance and cost records of previous projects.

A *materials and supplies office* is responsible for ensuring that the required construction materials, spare parts, and consumables are available in the correct quantity and at the right time for use by the maintenance work force. Figure 2.4 shows the organization of the Maintenance Control Portion of a large maintenance organization.

Maintenance Work Force

The maintenance work force is subdivided into several shops. All shops are overseen by a *maintenance supervisor, maintenance foreman*, or *general foreman*. The maintenance supervisor coordinates the various shops within the maintenance work force. Each of these shops is led by a *shop foreman* and consists of various craftsmen, apprentices, and helpers, the individuals who perform the actual work on the maintenance projects.

The maintenance operations of a large organization are very structured, due to the magnitude of work performed by the staff. Most work assignments should be well planned. They are assigned daily and usually require no modification during the course of the work day. However, supervision of the workers is important in maintaining a high level of productivity; the work force is apt to be somewhat unstable with a noticeable turnover rate among the helper and laborer categories of workers. For this reason, the training, retention, and promotion of employees needs constant monitoring. Figure 2.5 details the Work Force Branch of a large maintenance organization.

Intermediate Maintenance Organizations

The term *intermediate maintenance organization* applies to facilities such as shopping centers, community hospitals, small school districts, hotels, and small manufacturing plants. These facilities generally consist of one to four, medium to large physical structures with limited surrounding acreage. Intermediate facilities are usually fixed in size at initial construction, and thus have a limited need for modification or improvement projects. For example, a community hospital may undergo one or two major expansions and limited modifications in the entire lifetime of the facility as health-care technology changes dictate; a shopping center manager may occasionally reconfigure spaces to suit a new tenant; a school district may need to expand one of its schools to accommodate changed demography. Because such changes are infrequent and do not justify hiring an in-house engineering staff, these transient needs are handled by outside consultants. The facility manager, such as the hospital administrator or the superintendent of schools, is the point of contact and coordinator for outside engineering projects. A sample organizational chart depicting the structure of an intermediate maintenance organization is illustrated in Figure 2.6.

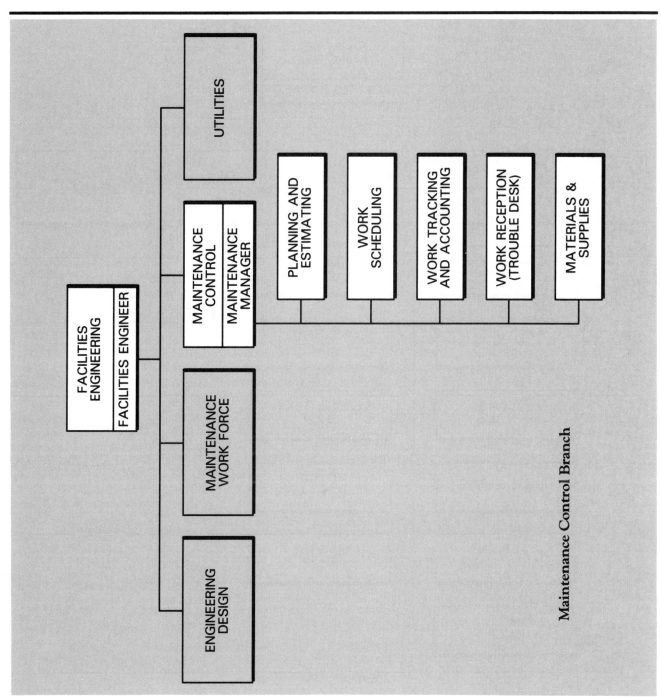

Figure 2.4

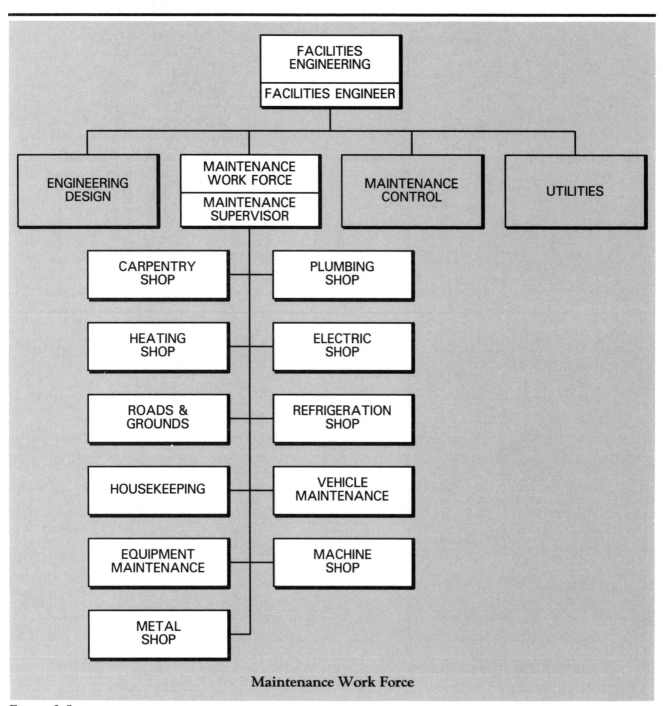

Maintenance Work Force

Figure 2.5

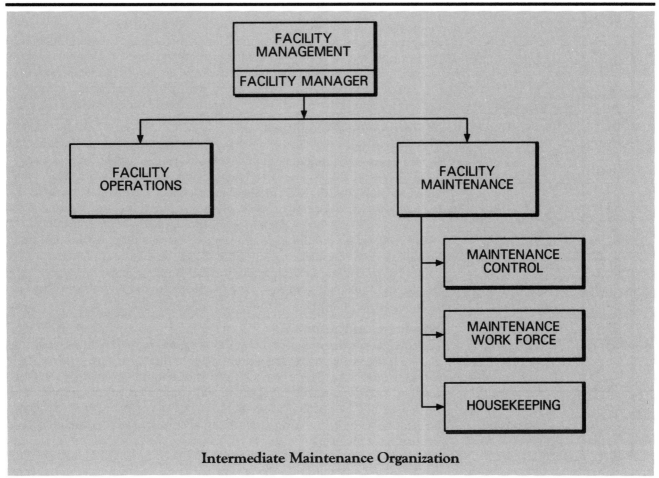

Intermediate Maintenance Organization

Figure 2.6

The basic elements of control and execution that apply to large maintenance organizations also exist within intermediate size organizations, but an intermediate facility's maintenance requirements do not support such a complex maintenance organization. All maintenance responsibility belongs to the maintenance supervisor. The maintenance supervisor is assisted by a small clerical staff, which handles the trouble desk, tracks workflow, and orders spare parts and consumables.

Maintenance Mechanics

The maintenance work force consists of a limited number of craftsmen formally described as *maintenance mechanics*, usually called *maintenance men*. These individuals are multi-talented and are able to work in a variety of disciplined areas. A single maintenance mechanic without a formal trade affiliation might work on minor plumbing, heating, and electrical problems. Since the various facility systems such as heating, plumbing, and air conditioning are limited, there is little justification for hiring full-time specialists to support each of these systems. Where sufficient system size or complexity justifies hiring a single full-time employee to maintain and repair a particular system, all work on that system is done in-house.

Service Contractors

For work other than minor troubleshooting and repair (such as replacing a leaking faucet or unclogging a drain) system difficulties are repaired by service contractors. These outside repairmen should be hired on retainer, familiar with the installed equipment, and committed to an appropriate response time. Depending upon the extent of use of contracted services, an intermediate organization employs an average of eight to twelve maintenance mechanics.

Housekeeping

Intermediate-sized facilities are often public buildings. Because the housekeeping effort is a significant proportion of the work, a formal housekeeping or janitorial staff is usually employed. In Figure 2.6, this group is designated being part of the maintenance branch. In facilities such as hotels, housekeeping requires a large staff and includes more than simply cleaning major public spaces. In such cases, the housekeeping staff is usually established as a separate entity working directly for the facility manager.

Organization

Operations in an intermediate maintenance organization are less structured than those of a large maintenance organization. More time is spent in emergency response, remedying minor discrepancies; little time is spent by the workers in modification and improvement projects, and, due to the contracting of services for major systems, little time is spent on classical preventive maintenance. Communication is a primary consideration for the intermediate maintenance organization. The key members of the maintenance staff usually carry electronic beepers, or are paged on a public address system. (See Chapter 7 for more information on communications.)

Small Maintenance Organizations

In small maintenance organizations, only one or two persons are involved in maintenance, in addition to a small housekeeping force. In such facilities, the maintenance man, a "jack of all trades," works directly for the facility manager. A small facility utilizes contracted services to a greater extent than intermediate organizations. In small facilities, major system components are repaired almost exclusively by service contractors.

Maintenance Personnel Attributes

Each of the maintenance personnel noted for the three different sizes of maintenance organizations has a specific set of assigned duties. Education or training is necessary for each position. Following are descriptions of each position, listing assigned duties and hiring criteria for each one.

Facility Manager

The overall duties of a *facility manager* depend on the objectives of the facility to be managed. The background of the facility manager matches the type of facility: a hotel manager has been trained in hotel management and a hospital administrator has studied health care administration. In both of these cases, the extent to which the formal college education programs address the function of facility maintenance is limited. In most fields, there is no formal exposure to facility maintenance. A newspaper publisher has extensive education in journalism and no training in running a printing plant, but is responsible for both functions. As a result, the facility manager is often not familiar with maintenance needs. The facility manager often views maintenance as a necessary evil: it seems to be a large cost item in the facility budget for which there is no identifiable product.

Facility managers may be either full or part owners of the facility, or may be salaried employees. The level of compensation varies considerably, depending primarily on the salary of counterparts within their respective industries. If not tied to an organization by ownership, facility managers are hired from outside the organization. If promoted from within the organization, facility managers are rarely, if ever, promoted from the maintenance branch of the organization.

Facilities Engineer

The *facilities engineer* is almost always a professional engineer. Training includes an engineering degree from a college or university. Civil, mechanical, industrial, and electrical engineering are traditional disciplines. Registration as a professional engineer in the state in which the facility is located is usually a requirement.

To obtain professional registration, the facility engineer must have graduated from a college or university which is accredited by the Accreditation Board for Engineering and Technology (ABET). ABET was formerly known as the Engineer's Council for Professional Development. ABET criteria establish minimum course offerings and course content for each of the engineering disciplines.

In addition to having obtained an accredited undergraduate degree, obtaining experience in the practice of engineering and passing a formal examination are necessary before a state license is granted. Most states require at least four years of experience in engineering before allowing an engineer to take the professional engineer's exam. The examination consists of two parts given separately. Part I, known as the Fundamentals, covers basic engineering theory as it is normally taught in the first three years of an undergraduate program. The Part I exam may be taken during the senior year of college. Once passed, an engineer is considered an engineer in training. For this reason, the Part I exam is often called the E.I.T. The second examination tests the applicant in engineering problems from a particular discipline. The exam is given over two four-hour sessions. During each of these sessions, the examinee has the choice of four out of several problems. The exam may be developed by the individual state, but most states utilize tests prepared by the National Council of Engineering Examiners (NCEE). Once an engineer is registered in one state, it is often possible to obtain registration in other states without retaking the formal examinations.

With the exception of a few emerging college degree programs in Public Works Management, there are no formal degree programs which address the field of facilities engineering. Similarly, the professional engineer's examination does not test the applicant on maintenance matters. The most closely related problems on the exam are in engineering economics.

Career paths to the position of facilities engineer are somewhat limited. In many organizations, few subordinate positions require an engineering degree; and since the position of facilities engineer is not an entry-level job, the necessary background must be obtained elsewhere. Most commonly, a facilities engineer has worked in a private consulting engineering firm, often designing the types of facilities and systems for which the facilities engineer is responsible.

Maintenance Supervisor

The *maintenance supervisor* (also called general foreman or superintendent) is a member of the facility management team with overall responsibility for coordinating the actual direct maintenance effort. The maintenance supervisor is usually a former craftsman who has progressed upward through the position of shop foreman to become the lead foreman in the organization. In addition to the formal training necessary for this position, the maintenance supervisor must have significant experience in supervising workers. Since familiarity with the facilities to be maintained is of prime importance to proper maintenance, the maintenance supervisor is generally a long-term employee of the organization, almost always promoted from within.

Planner/Estimator

The *planner/estimator* is a key member of the maintenance control team, responsible for accurately predicting the work crews and materials required to perform the wide variety of maintenance tasks. The planner/estimator must have a working knowledge of most general maintenance and construction processes. In order to properly perform the assigned duties, the planner/estimator must have personal knowledge of, or empirical data for, the factors listed below.

- Abilities of various maintenance craftsmen
- Traditional trade jurisdictions
- Proper craftsmen/apprentice/helper ratios
- Productivity rates for various maintenance tasks
- Equipment and tools required for various maintenance and construction tasks
- Material requirements for various maintenance tasks
- Material prices
- Normal use rate for consumable supplies
- Hourly labor costs for various maintenance workers
- Hourly operating costs for maintenance equipment

Generally, the planner/estimator acquires proficiency through on-the-job training. Accordingly, the planner/estimator is often a former maintenance craftsman with many years of experience. Since very limited formal training is available for this position, proper estimating is often more of an "art" than a "science." Historical data from within the maintenance organization and published cost data from estimating guides such as *Means Facilities Cost Data* are the most useful tools which aid the planner/estimator. The planner/estimator must have strong mathematical skills and be able to work at a desk for extended periods of time. Therefore, not all maintenance craftsmen are likely to be good candidates for the job of planner/estimator.

In addition to constantly producing reliable estimates of manpower, equipment, material, and spare parts requirements for various tasks, the planner/estimator must also continually review the accuracy of those estimates. Computerized maintenance management systems, while not replacing the need for a planner/estimator, are making the job of producing accurate estimates easier. Such systems allow the planner/estimator to continually compare the estimates with actual expenditures, and allow for long-term comparisons of similar work items to determine trends or averages.

Scheduler

The *scheduler* is a member of the maintenance management team of a large maintenance organization. The scheduler coordinates the efforts of multi-disciplinary maintenance projects. When projects, generally improvement or replacement related, require craftsmen from two or more shops, it is necessary to divert these craftsmen from their routine maintenance work at a common time to complete the project. The scheduler works with the estimated man-hour requirements produced by the planner/estimator and with the available resources from the appropriate shops to schedule these joint effort projects. In order to accomplish this, the scheduler must maintain familiarity with the normal routines of each of the maintenance shops in order to determine the amount of time each shop has available to devote to major multi-disciplinary projects.

Projects are scheduled in accordance with their established priority to fully utilize all personnel without disturbing the routine preventive maintenance activities. The scheduler is responsible for situations where the project involves the use of outside contractors for some portion of the work, and when significant material deliveries must be coordinated.

In addition to scheduling the individual participating shop personnel, the scheduler coordinates the total project within the routine operations of the facility. Therefore, the scheduler must be familiar with the facility operations and must consider the impact of any disruptions that accompany each type of project. To accomplish this coordination, the scheduler is frequently in direct contact with the facility users before a project is scheduled and later during project execution. The activities of the scheduler have a direct impact on customer and user satisfaction with overall facilities maintenance.

In companies with few improvement projects and, therefore, a minimal need to use several different trades on individual projects, the duties of the scheduler are absorbed by the planner/estimator. In those cases, the scheduling of multiple trades is arranged by the various shop foremen involved.

Shop Foreman

The *shop foreman* is the designated leader of a group of maintenance mechanics or craftsmen and is a critical individual within a maintenance organization. The shop foreman is the final point of contact for conveying management decisions, assignments, and priorities to the maintenance personnel performing the work.

Since shops are most frequently divided by trade, the shop foreman is almost always a former craftsman of that specific trade. Carpenters become carpenter shop foremen, plumbers or pipefitters become plumbing shop foremen. Experience, however, while very helpful, is not the only qualification to consider when selecting a shop foreman. The primary role of the shop foreman is to direct the efforts of the craftsmen, apprentices, and helpers within the shop. Therefore, strong leadership skills are mandatory. In addition, as the craftsman moves toward the position of shop foreman, the amount of paperwork increases significantly. The shop foreman must be able to complete the various reports, evaluations, requisitions, and inventories associated with the day-to-day operation of the shop.

Generally, the foreman is promoted from the pool of craftsmen within the organization. However, the selection of a foreman should be carefully considered. The most popular worker may not be sufficiently dedicated to organizational goals; the best leader may lack administrative skills; and the craftsman with administrative skills may not have the respect of fellow workers. The interviewer for the position should emphasize to the candidate the organizational goals and administrative requirements of the position. In addition, the maintenance supervisor should monitor the performance of a new foreman. Careful selection and closely monitored development of shop foremen results in lower maintenance costs.

Maintenance Craftsmen

The positions of facilities engineer, facilities maintenance manager, planner/estimator, scheduler, and shop foreman exist to identify and schedule maintenance work. The actual completion of the maintenance work, however, lies in the hands of the maintenance craftsman or laborer.

Numerous trades make up the pool of maintenance craftsmen. Large maintenance organizations employ full-time individuals from a variety of different trades, while smaller organizations require individuals with talents which transcend traditional trade jurisdictions. A large maintenance organization may have teams of millwrights, carpenters, cabinet makers, and plasterers; a small organization might rely upon a maintenance mechanic or carpenter to perform these various functions. There is an obvious loss in quality of workmanship when a single individual is called upon to perform the work of various trades. However, the degree to which an organization enjoys increased productivity from specialists is strictly a function of need. Only where there is an identified need for the full-time employment of any individual specialized trade should one or more craftsmen of that trade exist in a maintenance organization.

The following is a list of the various common trades which exist in facilities maintenance organizations:

- Carpenters
- Millwrights
- Cabinet makers
- Locksmiths
- Glazers
- Roofers
- Plasterers
- Welders
- Sheet Metal Workers
- Pipe Fitters
- Plumbers
- Machinists
- Auto Mechanics
- Diesel Mechanics
- Oilers
- Electricians
- High Voltage Electricians
- Instrument Repair Technicians
- Splicers
- Air Conditioning and Refrigeration Mechanics
- Equipment Operators
- Operating Engineers
- Stationary Engineers
- Boilermakers
- Riggers
- Painters
- Sign Makers
- Janitors
- Laborers
- Helpers (for all trades)
- Apprentices (of all trades)
- Maintenance Mechanics

The above list is by no means all inclusive. Also, the titles are not necessarily the only ones recognized for each maintenance trade specialty. The exact title for a person who performs a common set of tasks is determined either by the organization or through the local practice within a geographic region. The titles are generally dictated by state licensing laws and regulations.

Most craftsmen obtain their positions through a combination of technical training and on-the-job experience. Technical training in each of the listed trades is available through both vocational programs in high school and through privately run trade and technical schools. Courses are offered at night through many public school systems to prepare students for the various trades. Particular emphasis in these night programs is placed on trades for which formal licensing requirements exist. These licensing requirements are generally established by a state to ensure a uniform level of knowledge and experience for trades which directly affect the safety of the public. The acquisition of a license may require completion of a minimum number of hours of classroom instruction, the passing of a formal exam, a minimum period of apprenticeship under a licensed craftsman, or a combination of the three. Apprentice programs are generally established and monitored by either the state or the associated labor union. An apprenticeship consists of several years of progressively more responsible work experience conducted under the direct supervision of a craftsman of the same trade.

One prevalent source of training and experience for craftsmen is through service in the military. Each branch of the military maintains a set of formal schools and correspondence courses in most of the trades previously listed. The graduates of these schools and courses then work in that trade area, acquiring extensive experience comparable to private-sector facilities maintenance work experience. Each of the branches of the military has numerous large bases, each with extensive maintenance responsibilities and well-defined maintenance organizations.

The Maintenance Mechanic

The title of *maintenance mechanic* applies to an employee whose duties encompass, to a certain extent, many of the individual trades on the previous list. The extent to which an individual maintenance mechanic carries out the duties of several trades is decided by the facilities maintenance manager. Common sense should be applied when combining several responsibilities under one single job title. For example, few individuals are sufficiently proficient as both high voltage electricians and cabinet makers.

Combining several trade skills under the title of maintenance mechanic is common practice in small maintenance organizations which cannot support full-time specialists. The work associated with a "handyman" or apartment building superintendent is an extreme example of the duties of a maintenance mechanic. In most cases, the maintenance mechanic is responsible for relatively easy tasks which transcend several trade boundaries. A typical maintenance mechanic might be expected to perform the tasks listed below.

- Unclog drains, replace leaky washers, and other minor plumbing repairs.
- Rehang doors and adjust locks and door hardware.
- Replace broken panes of glass.
- Patch small holes in plaster and drywall.
- Patch concrete paving.
- Adjust heating controls.
- Replace floor or wall tiles.

While it is possible for craftsmen from several different trades to perform each of the above, none of these tasks requires the highest level of qualifications or proficiency from its most closely associated trade. In intermediate maintenance organizations, the maintenance mechanic is likely to be a person with more limited duties, perhaps being responsible for two or three trade disciplines for which there is insufficient need to justify a single employee. For example, an organization might combine welding and sheetmetal work; plastering and painting; or glazing, door, window, and lock repair.

Since there is generally sufficient work in large maintenance organizations to justify the full-time employment of one or more individuals of each trade discipline, the main reason for utilizing a maintenance mechanic is to allow the use of single person crews. The relocation of a door within a facility could conceivably require the work of a carpenter, plasterer, locksmith, and painter. The coordination of these four persons, each contributing one or two hours to the project, is likely to be difficult, resulting in wasted time if the job is not ready for the next successive trademan. For a door relocation job, a maintenance mechanic could remove the door, relocate the frame, patch the walls, and then reinstall the door.

From the journeyman level in one trade discipline, frequently carpentry, an individual, through training or life experience, becomes sufficiently proficient at the basic level of other trades to assume the expanded duties of a maintenance mechanic. A maintenance mechanic in a small organization usually develops proficiency in minor repairs to the various systems of that facility. However, these skills are often limited to those specific systems; maintenance mechanic experience in one facility does not imply equal proficiency in others, even if they are of similar complexity. For this reason, great care must be taken in the selection of a new maintenance mechanic.

The maintenance mechanic usually needs a "break-in" period in order to become familiar with new systems. The facility maintenance manager should expect this orientation period to be somewhat longer than that of a craftsman in a single discipline, whose orientation to a new facility is likely to be shorter due to the already intensive training and experience within a specific field.

Maintenance Personnel Management

The bottom line in personnel management is *productivity*. Employing individuals with the requisite skills is not sufficient to ensure an appropriate level of facility maintenance productivity. The facility maintenance manager must use sound techniques of personnel management and development to maximize the productivity of the entire maintenance work force. The personnel management principles that have the greatest effect on productivity are listed below.

- Definition of position duties and responsibilities
- Evaluation of performance
- Training of employees

Both employer and employee must fully understand the duties associated with each job. The performance of the individuals must be measured so that unacceptable levels of performance can be corrected. And the proficiency of employees must be maintained and developed. These three personnel management principles are discussed in the following sections.

Defining Maintenance Positions

The role of each individual within a maintenance organization must be clearly defined. A formal job description is necessary to ensure that each employee is cognizant of the assigned duties and the standards against which performance will be measured. A job description is a statement of the duties of a position and the basic skills expected of the employee, and establishes work rules by which to govern employee behavior. Lack of accurate job descriptions for each maintenance position leads to confusion and improper worker performance. Further, salient items from the job description are extracted and paraphrased when advertising for the position. If particular factors such as physical effort required or overnight travel are a condition of employment, they should also be included within the job description.

As previously noted, the title assigned to a position is insufficient to define the varied duties of the position. Because each organization has different expectations of its employees, job descriptions define the what, where, when, and how of each position within a particular maintenance organization. The compilation of all job descriptions within an organization describe the total maintenance effort.

For many positions there is a formal licensing process which establishes minimum levels of performance and knowledge for the practice of that trade. In cases where possession of a license is a requirement for employment, the need for describing the wide variety of basic skills and duties necessary for the position is reduced, but not eliminated. For positions that do not require the possession of a license, the required skills and duties for the job must be described in detail. For example, the statement "Locate and repair leaks in water and waste piping" implies that the person selected to fill that position must have the following basic skills:

- Cut and solder copper pipe.
- Cut threads on iron pipe.
- Connect threaded pipe.
- Rebuild, repair, or replace a variety of valves.
- Renew joints in waste piping systems.
- Trace piping systems back to the source of a leak.
- Identify valves necessary to isolate a section of pipe.

The statement above implies that a variety of basic plumbing and pipefitting skills are necessary for the position. In addition to making some general statements about the duties of a job, specific systems of the facility or special areas of emphasis within the trade must be included in the description. For example, the statement "Repair or rebuild steam pressure-reducing valves" describes a specific duty.

The majority of the statements of duties concern manual tasks. When describing maintenance positions, however, it is important to describe mental activities. Since many maintenance projects involve malfunctioning systems or equipment, the ability to diagnose a problem and to determine the scope of repairs is necessary for many maintenance positions. The worker may have to study equipment manuals and engineering drawings in order to properly diagnose a system malfunction. The statement "Diagnose and repair refrigeration system malfunctions" implies knowledge of various systems and their component parts, deductive reasoning ability, and manual skills necessary to repair different refrigeration systems. Figure 2.7 is a position description for an air conditioning and refrigeration mechanic.

Assistance and Supervision: After describing in detail the duties of the position, the next step is to describe *how* the work is to be performed. Two factors are included in this part of the job description: the degree of assistance available to the worker, and the level and degree of supervision. If the job requires that the worker perform non-dangerous tasks without additional assistance, it must be stated. For example, the statement "Most work is completed independently" implies that the worker is not dependent upon apprentice or laborer help. The concept of independent work is often at odds with union work rules and trade jurisdictions. Where union contracts allow or where the work force is not unionized, the statement above greatly enhances productivity. A worker also receives assistance through the use of various tools and diagnostic equipment. However, labor-saving equipment is often expensive and usually requires some specific training. (See Chapter 5 for more information on the decision to purchase labor-saving devices.)

JOB DESCRIPTION

POSITION TITLE Air conditioning and Refrigeration Mechanic

PLACE OF WORK

Employee works throughout the facility on various equipment. Employee also performs work in the air conditioning shop.

DUTIES AND RESPONSIBILITIES:

1. Routine servicing of installed air conditioning systems and components including: compressors, condensers, valves, coils, cooling towers, heat pumps, motors, belts, bearings, fans, filters, piping, wiring, control systems, thermostats, ducting, ventilators, and other appurtenances.
2. Performs routine and emergency repairs to same.
3. Troubleshoots and diagnoses malfunctioning systems and components.
4. Interprets engineering drawings and equipment manuals related to air conditioning systems and component parts.
5. Performs replacement of major and minor parts, components, or entire systems.
6. Performs overhauls on major equipment components.
7. Provides recommendations regarding air conditioning systems maintenance and repair.
8. Performs other duties as assigned.

ASSISTANCE PROVIDED

Generally works independently. Occasionally as member or leader of a repair team. Uses hand tools, machines, diagnostic equipment. Reads and interprets equipment manuals.

ASSIGNED SUPERVISOR

Works directly for the Foreman, Air Conditioning Shop.

SUPERVISORY RESPONSIBILITY

Occasionally supervises apprentice or journeymen mechanics during repairs. Occasionally supervises laborers.

WORK ENVIRONMENT

Work is performed indoors and outdoors. Indoor work is done in cramped or crowded spaces. Spaces are not usually air conditioned. Work is performed on or near operating equipment. Outside work is performed in all climates, often during inclement weather. Work at ground level and on roofs.

HOURS OF WORK

Normal hours of work are Monday thru Friday, 7:30 AM to 5:00 P.M. Occasional overtime. Occasional holiday work. Frequent after hours calls for emergency repair work.

Figure 2.7

If a maintenance organization uses a specific brand of equipment, state the name of the equipment. These equipment-specific statements are often included as part of the primary list of duties such as, "Enter and update data and generate reports using MAINTAINS computer systems."

In addition to the assistance available to the worker, the second factor to consider when describing how a job is to be performed is the level and degree of supervision provided and the degree to which the worker is required to supervise others. Some sample job description statements regarding supervision are listed below:

- Work is performed under the direct supervision of the shop foreman.
- Work is performed without direct supervision.
- Worker supervises work crews of three to five craftsmen.
- Worker occasionally directs craftsmen from several trades and shops in large repair projects.

When describing a position which is primarily supervisory in nature, the above statements are limited to defining the persons to whom the worker reports. The specific supervisory duties are spelled out in the statement of duties portion of the job description.

When describing supervisory positions, the degree of personal authority must be delineated. Personal authority is the degree to which the worker has final authority to commit the maintenance resources of money, men, and equipment in the course of their job performance. Figure 2.8 continues the development of a job description.

Work Environment and Location: The job description also defines the places in which the worker will be employed. This is important, since it is often necessary for a worker to be employed out of doors, in several buildings, at various locations throughout a facility complex, or in a dangerous environment. Clarity and mutual understanding is enhanced if such conditions of employment are included within the job description.

The worker should be fully aware of the conditions of employment when initially hired. Should a worker claim that certain assignments are beyond the scope of their job description, management must be able to cite the original conditions of employment. A clear job description backs up the conditions under which an employee was hired. For this reason, the preferred order for establishing a new position is first to write the job description, and then to extract from it for hiring purposes.

Much maintenance work is carried out in confined spaces, on ladders, on roofs, and in all types of weather. Further, many maintenance tasks must be performed in environments where special safety practices must be applied. The use of hard hats, eye protection, respirators, or special clothing may be required. While specific reference to such work conditions is intended to convey requirements to an employee, it also implies certain responsibilities to the employer. The statement, "Service aircraft warning lights on the building antenna tower using furnished climbing safety equipment," describes a worker duty, a work environment, and an employer responsibility. Figure 2.9 shows specific statements in the work environment portion of the example job description.

JOB DESCRIPTION

POSITION TITLE Air conditioning and Refrigeration Mechanic

PLACE OF WORK

Employee works throughout the facility on various equipment. Employee also performs work in the air conditioning shop.

DUTIES AND RESPONSIBILITIES:

1. Routine servicing of installed air conditioning systems and components including: compressors, condensers, valves, coils, cooling towers, heat pumps, motors, belts, bearings, fans, filters, piping, wiring, control systems, thermostats, ducting, ventilators, and other appurtenances.
2. Performs routine and emergency repairs to same.
3. Troubleshoots and diagnoses malfunctioning systems and components.
4. Interprets engineering drawings and equipment manuals related to air conditioning systems and component parts.
5. Performs replacement of major and minor parts, components, or entire systems.
6. Performs overhauls on major equipment components.
7. Provides recommendations regarding air conditioning systems maintenance and repair.
8. Performs other duties as assigned.

ASSISTANCE PROVIDED

Generally works independently. Occasionally as member or leader of a repair team. Uses hand tools, machines, diagnostic equipment. Reads and interprets equipment manuals.

ASSIGNED SUPERVISOR

Works directly for the Foreman, Air Conditioning Shop.

SUPERVISORY RESPONSIBILITY

Occasionally supervises apprentice or journeymen mechanics during repairs. Occasionally supervises laborers.

WORK ENVIRONMENT

Work is performed indoors and outdoors. Indoor work is done in cramped or crowded spaces. Spaces are not usually air conditioned. Work is performed on or near operating equipment. Outside work is performed in all climates, often during inclement weather. Work at ground level and on roofs.

HOURS OF WORK

Normal hours of work are Monday thru Friday, 7:30 AM to 5:00 P.M. Occasional overtime. Occasional holiday work. Frequent after hours calls for emergency repair work.

Figure 2.8

JOB DESCRIPTION

POSITION TITLE Air conditioning and Refrigeration Mechanic
PLACE OF WORK
Employee works throughout the facility on various equipment.
Employee also performs work in the air conditioning shop.
DUTIES AND RESPONSIBILITIES:

1. Routine servicing of installed air conditioning systems and components including: compressors, condensers, valves, coils, cooling towers, heat pumps, motors, belts, bearings, fans, filters, piping, wiring, control systems, thermostats, ducting, ventilators, and other appurtenances.
2. Performs routine and emergency repairs to same.
3. Troubleshoots and diagnoses malfunctioning systems and components.
4. Interprets engineering drawings and equipment manuals related to air conditioning systems and component parts.
5. Performs replacement of major and minor parts, components, or entire systems.
6. Performs overhauls on major equipment components.
7. Provides recommendations regarding air conditioning systems maintenance and repair.
8. Performs other duties as assigned.

ASSISTANCE PROVIDED
Generally works independently. Occasionally as member or leader of a repair team. Uses hand tools, machines, diagnostic equipment. Reads and interprets equipment manuals.

ASSIGNED SUPERVISOR
Works directly for the Foreman, Air Conditioning Shop.

SUPERVISORY RESPONSIBILITY
Occasionally supervises apprentice or journeymen mechanics during repairs. Occasionally supervises laborers.

WORK ENVIRONMENT
Work is performed indoors and outdoors. Indoor work is done in cramped or crowded spaces. Spaces are not usually air conditioned. Work is performed on or near operating equipment. Outside work is performed in all climates, often during inclement weather. Work at ground level and on roofs.

HOURS OF WORK
Normal hours of work are Monday thru Friday, 7:30 AM to 5:00 P.M. Occasional overtime. Occasional holiday work. Frequent after hours calls for emergency repair work.

Figure 2.9

Hours of Work: In addition to describing what, how, and where the work is to take place, the job description should also define the formal hours of work. Many organizations have written policies that address the standard hours of operation for a facility and the standard work hours for different classifications of employees. The job description, in addition to being consistent with these published rules, should also address the frequency of overtime, the frequency of assignment to various work shifts, and any requirements concerning being recalled to the workplace for emergency work or repairs. Figure 2.10 is a completed job description.

Performance Appraisal

Another critical element in personnel management is evaluating the performance of the employee. The individual worker is the most variable factor in facility maintenance productivity. A performance appraisal system should be designed to sustain or improve a high level of worker productivity. The elements of a performance appraisal system are listed below.

- Tasks to be evaluated
- Standards of performance
- Standard appraisal form
- Schedule of frequency of evaluation
- Responsibility for completing the evaluation
- Rewards and sanctions for satisfactory and unsatisfactory appraisals
- Review and monitoring of the appraisal system

Each of these points is addressed in the following section.

Tasks to be Evaluated: The job description is a statement of the duties and obligations of the employee. While it is desirable to evaluate employee performance in all of the areas of the job description, the sheer volume of such an appraisal process would overwhelm a supervisor. In addition, the job description does not address the quality of performance; it implies absolute performance. The appraisal should reflect how well the employee performed those duties originally outlined in the job description.

Instead of an item-by-item evaluation, an appraisal system should separate an employee's duties into categories or areas of emphasis. The areas are defined by the quality of the employee's performance in each task area to be assessed. For example, a job description for the the position of receptionist lists the specific tasks involved: answer the phone, receive requests from facility users for maintenance assistance, enter work requests in the maintenance computer system, type office correspondence, and maintain the work request files. An appraisal for the same position evaluates the quality of the tasks performed: accuracy of work, timeliness of work, and phone manners. The subjective evaluation of performance quality is usually quantified by grading or classifying the performance into categories, such as outstanding, excellent, above average, average, or unacceptable.

These areas of emphasis form the basis for goal setting by both the employer and the employee. The employee should be fully aware of the areas of emphasis at the start of each evaluation period. Figure 2.11 shows some typical areas of emphasis, or goals for maintenance employees.

JOB DESCRIPTION

POSITION TITLE Air conditioning and Refrigeration Mechanic

PLACE OF WORK

Employee works throughout the facility on various equipment. Employee also performs work in the air conditioning shop.

DUTIES AND RESPONSIBILITIES:

1. Routine servicing of installed air conditioning systems and components including: compressors, condensers, valves, coils, cooling towers, heat pumps, motors, belts, bearings, fans, filters, piping, wiring, control systems, thermostats, ducting, ventilators, and other appurtenances.
2. Performs routine and emergency repairs to same.
3. Troubleshoots and diagnoses malfunctioning systems and components.
4. Interprets engineering drawings and equipment manuals related to air conditioning systems and component parts.
5. Performs replacement of major and minor parts, components, or entire systems.
6. Performs overhauls on major equipment components.
7. Provides recommendations regarding air conditioning systems maintenance and repair.
8. Performs other duties as assigned.

ASSISTANCE PROVIDED

Generally works independently. Occasionally as member or leader of a repair team. Uses hand tools, machines, diagnostic equipment. Reads and interprets equipment manuals.

ASSIGNED SUPERVISOR

Works directly for the Foreman, Air Conditioning Shop.

SUPERVISORY RESPONSIBILITY

Occasionally supervises apprentice or journeymen mechanics during repairs. Occasionally supervises laborers.

WORK ENVIRONMENT

Work is performed indoors and outdoors. Indoor work is done in cramped or crowded spaces. Spaces are not usually air conditioned. Work is performed on or near operating equipment. Outside work is performed in all climates, often during inclement weather. Work at ground level and on roofs.

HOURS OF WORK

Normal hours of work are Monday thru Friday, 7:30 AM to 5:00 P.M. Occasional overtime. Occasional holiday work. Frequent after hours calls for emergency repair work.

Figure 2.10

Typical Management Goals

Maintenance Managers
(Facilities Engineer, Facility Maintenance Manager, etc.)

Provide timely maintenance and repair response
Identify cost saving maintenance methods
Reduce maintenance related down-time
Operate maintenance program within budget
Coordinate improvements to reduce operating costs

Maintenance Control Personnel
(Planner/Estimator, Scheduler, Work Receptionist, etc.)

Minimize lost time due to lack of proper materials
Produce reasonable estimates
Minimize lost time due to poorly coordinated projects
Produce workable preventive maintenance work schedules
Provide courteous service to customers
Minimize plant/facility down-time due to maintenance projects
Accurately describe problems on work orders
Maintain up-to-date maintenance files
Coordinate expenditures against budgeted amounts

Maintenance Supervision
(General Superintendent, Shop Foreman, etc.)

Execute projects within estimated time and cost
Ensure full utilization of employees
Maintain accurate spare parts and materials inventories
Plan purchases well in advance
Improve shop productivity
Minimize small tool losses
Identify needed replacement projects
Maintain accurate equipment/machinery history files

Maintenance Workers
(Craftsmen, Laborers, Janitors, Helpers)

Complete projects thoroughly
Keep supervisor advised of project work status
Maintain cleanliness around work sites
Treat customers courteously
Identify potential equipment failures
Observe proper safety practices

Figure 2.11

Note that the goals vary considerably in character from the maintenance craftsman up to the facilities engineer. The breadth of responsibility of the facilities engineer precludes listing specific areas of emphasis for individual daily maintenance projects. Rather, the facilities engineer's areas of emphasis are more aligned with the overall goals of the maintenance organization and the firm that it serves. The craftsman, however, deals with small projects, or portions of small projects. The areas of emphasis, therefore, can be limited to the desired characteristics of those projects and tasks.

Standards of Performance: Performance appraisal systems may utilize a multi-step scale from unsatisfactory to outstanding or may be limited to general satisfactory/unsatisfactory ratings. Either way, the employee needs to know what level of performance earns each of the available ratings. All established standards of performance should be directly tied to the concept of productivity.

In many cases, the task evaluated can be quantitatively assessed. For example, the proficiency of a typist is measured by the number of typing errors; a plant manager's performance is measured by units of production; a salesman is rated by total sales. Within the maintenance workforce, there are numerous opportunities for quantitative evaluation where the standards of performance can be objectively established. Figure 2.12 lists typical quantitative standards of performance for the various maintenance employees.

In other cases, the ability to quantitatively evaluate performance is difficult, and a subjective judgment is necessary. A receptionist must be polite, but that trait is difficult to break down into various grades. For an accountant, the books either balance or they don't, but the ease of tracing an individual transaction is evaluated subjectively. Figure 2.13 lists typical subjective standards.

Completing the Appraisal: Once the tasks to be evaluated and their respective standards of performance are defined, the completion of the appraisal is relatively simple. The evaluator, relying upon records, notes, and recollection, assigns an appropriate rating for each area of emphasis. Figures 2.14 and 2.15 are typical performance appraisal reports for two categories of maintenance employees. Two different formats are shown, one for personnel involved in hands-on direct or indirect maintenance work, and one for those involved in supervisory maintenance positions.

Using the Appraisal: The underlying objective of an evaluation system is to *maintain or increase productivity* of the maintenance workforce. In order to achieve this objective, the employees must be advised of the results of the most recent appraisal and must be counseled concerning what actions they may take to earn a higher rating. A few employees are satisfied with minimal performance. Most employees, however, desire to do the best job that they can, because they take pride in themselves and their work and want to be recognized for their performance. The appraisal system is a formal method for providing that recognition.

In addition to the personal satisfaction that a positive evaluation provides, an appraisal system must include specific rewards and sanctions. An extreme example for an unsatisfactory appraisal rating is dismissal; an outstanding rating yields possibly a raise, a bonus, or even a promotion. If the organization does not intend to use the appraisal as a basis for such reward or punishment, then the time involved in designing the appraisal system, developing the standards, and making the evaluations is used better elsewhere. If the appraisal system does not maintain or improve production, then the time involved subtracts significantly from the productivity of all the individuals involved.

For a very small maintenance organization, a verbal reward or reprimand, as appropriate, is a sufficient performance appraisal system. Where the benefits to be derived from an appraisal system are less than the costs of establishing and administering the system, at the very least the maintenance manager should establish formally the areas of emphasis for each employee. This minimal effort provides each employee with a better understanding of management desires and organizational goals.

Quantitative Standards of Performance

Maintenance Managers
(Facilities Engineer, Facility Maintenance Manager, etc.)

Reduce budget by 15%
Reduce plant/facility down-time by 30%
Maintain 15 minute response time for emergencies
Projects to be completed within $+/-10\%$ of budget

Maintenance Control Personnel
(Planner/Estimator, Scheduler, Work Receptionist, etc.)

Maintain accuracy of $+/-5\%$ for material cost estimates
95% of all projects coordinated with no lost time
Work orders filed within 3 days of completion
No more than 5 work order backlog for entry in computer
Accounts to be kept in balance, reconciled weekly

Maintenance Supervisors
(General Superintendent, Shop Foreman, etc.)

Complete 90% of preventive maintenance work on schedule
Maintain 2 week supply of consumables
Forward all completed project reports weekly
Identify potential budget variances in advance
Maintain accurate inventory of spare parts
Work to be entered on equipment history within 3 days

Maintenance Workers
(Craftsmen, Laborers, Janitors, Helpers)

Complete work within budgeted time
Clean 20,000 square feet daily
Maintain full levels of towels, toilet paper in rest rooms
Perform six service calls per hour
Recharge air conditioning unit in two hours
Service vehicles within 15 minutes of request
Respond to emergency calls within 2 minutes of paging

Figure 2.12

If the costs and benefits of establishing an appraisal system indicate that the system will be cost effective, the next step in the process is to establish the frequency of evaluation for each level of employee. No set rules apply to frequency; as a general guideline, annual evaluations are usually sufficient for craftsmen, since their areas of emphasis rarely change. For management employees, a semi-annual evaluation is desirable, since the areas of emphasis are more subject to change. Changes in management personnel areas of emphasis are made as previous areas are satisfied or as organizational goals change.

Training for Maintenance Personnel

The initial hiring of an employee involves finding an applicant whose qualifications most closely match the demands of the defined position. Because even the best applicant rarely possesses *every* quality and skill required for the job, immediate training needs include adaptation to the procedures of the maintenance organization, the particular tools used for

Subjective Standards of Performance

Maintenance Managers
(Facilities Engineer, Facility Maintenance Manager, etc.)

Maintain courteous customer relations
Identify potential cost savings
Plan for emergencies
Identify potential system failures
Perform a "proper" preventive maintenance program

Maintenance Control Personnel
(Planner/Estimator, Scheduler, Work Receptionist, etc.)

Maintain courteous treatment of customers
Identify shop procedural problems
Identify maintenance trends
Maintain orderly maintenance project files

Maintenance Supervisors
(General Superintendent, Shop Foreman, etc.)

Reduce employee tardiness
Schedule workers efficiently

Maintenance Workers
(Craftsmen, Laborers, Janitors, Helpers)

Treat customers courteously
Identify potential methods for reducing maintenance costs
Uniforms to be neatly pressed
Carpets to be thoroughly vacuumed

Figure 2.13

EMPLOYEE PERFORMANCE APPRAISAL

Position Title _____ Incumbent _____

	Appraisal	
	From	To

Areas of Emphasis

Evaluation of Performance

O-Outstanding E-Excellent S-Satisfactory U-Unsatisfactory

Specific Performance Factor Evaluated	Rating
Demonstrated performance of basic duties	
Performance in areas of emphasis	
Accepting direction	
Working independently	
Personal appearance, job cleanliness	
Ability to work effectively with other trades (teamwork)	
Timeliness, attendance	
OVERALL PERFORMANCE RATING_____	
Areas needing improvement:	

Appraisal completed by:

Name:_____Title_____Date_____

Employee acknowledgement_____Date_____

Figure 2.14

47

SUPERVISORY EMPLOYEE PERFORMANCE APPRAISAL

Position Title	Incumbent	Appraisal From	To

Areas of Emphasis

Evaluation of Performance
O-Outstanding E-Excellent S-Satisfactory U-Unsatisfactory

Specific Performance Factor Evaluated	Rating
Demonstrated performance of basic duties	
Ability to lead workers to improved performance	
Ability to determine goals, priorities	
Communications abilities (written & oral)	
Cost consciousness	
Ability to work effectively with other staffs	
Support and enforcement of policy	
OVERALL PERFORMANCE RATING_____	

Areas needing improvement:

Appraisal completed by:

Name:_____Title_____Date_____

Employee acknowledgement_____Date_____

Figure 2.15

48

maintenance performance, and the specific facilities and systems to be maintained. Reasons for training maintenance employees are listed below.

- Maintaining or improving proficiency
- Adapting to new maintenance equipment or methods
- Response to observed shortfalls in performance
- Legislative mandate
- Improved employee relations
- Loss prevention

Proficiency training for most employee tasks is limited. Generally, proficiency is maintained by the day-to-day practice of the job or trade. Tasks that are performed less frequently, however, may mandate occasional training. Training is often provided as insurance against future needs. A welder might take a refresher course in the welding of aluminum to make up for a lack of hands-on practice. A facilities engineer might take a course in construction contract claims, if the position infrequently involves construction contracts.

In many instances, changes in the facility or equipment require that various members of the maintenance work force receive formal training. For example, a change in a heating or ventilating control system, the installation of a new fire or intrusion alarm, or the purchase of a different vehicle requires that previously proficient craftsmen work on unfamiliar equipment. The costs of such training must be included in any cost to benefit analysis that addresses changes of these types.

Changes in maintenance management methods require training. For example, the adoption of a new or replacement computer system for managing maintenance requires training both the computer operators and the personnel who work with the computer-generated work orders. A change in management policy concerning how maintenance problems are reported dictates training for the facility employees not involved in maintenance work performance.

The constant observation of ongoing maintenance activities may reveal a shortfall in the job performance of an individual or group of individuals. Training to remind employees of the procedures of the organization or the methods of maintaining the facility or equipment is a way of maintaining procedure discipline. If frequent complaints arise from facility users that project sites are left unclean or that workers behavior is inappropriate, training is necessary for all related personnel. Such training is usually provided by the first line supervisor.

Changes in and additions to the legislative codes and regulations that govern the conduct of a facility mandate immediate training. For example, when the dangers of asbestos insulation were discovered, a complex system of safeguards evolved, requiring significant training on asbestos identification, encapsulation, removal, and disposal. Recent major problems related to the improper disposal of hazardous waste have led to a complex set of regulations and methods for tracking the handling and disposal of all hazardous waste. In addition to the ethical considerations of the improper disposal of hazardous waste, the implications in terms of cost to a facility discovered improperly disposing of hazardous waste is sufficient justification for extensive and frequent training. In terms of knowing the law, employees in leadership positions also need to have an understanding of legislated personnel issues, such as racial and sexual discrimination and sexual harassment.

Since a lack of any of the above types of training could lead to serious financial losses for the facility, providing necessary training is a form of loss prevention. Loss prevention training is a term most frequently applied to safety-related training. Personnel must be continually reminded about proper job-related safety procedures, and also must be taught procedures for dealing with emergencies. Often, courses are taught in first aid and cardiopulmonary resuscitation (CPR). In response to the long-term employee absences associated with back injuries, many companies have instituted courses to teach personnel the proper procedures for lifting heavy objects. In short, training should be provided whenever potential loss justifies the cost of that training.

Sources of Training

Training is not limited to formal classroom lectures and demonstrations. Much informal training is obtained through the personal efforts of the individual workers. There are many maintenance-related organizations, and membership in such organizations normally entitles the member to various trade publications outlining current industry trends. Trade magazines and special interest journals are perhaps the cheapest form of training available. An employee may discover a new maintenance procedure or tool in a trade magazine. Textbooks have a similar role and potential as training tools.

Generally, training is performed by members of the maintenance organization or by representatives from within the parent organization. There are several advantages to in-house training: the costs for training are significantly reduced; the training is tailored to the specific facility or system; and the scheduling of the training is much more flexible. Despite these benefits, however, in-house training often ends with unsatisfactory results, because the person delivering the training is an expert in the field, but has limited teaching ability or experience. The alternatives to in-house training are courses and seminars taught by commercial sources or maintenance-related trade and professional organizations. Seminars are sponsored by local trade schools, colleges, and national professional organizations or local chapters of those organizations, and are also conducted in conjunction with trade shows or exhibitions. The availability of courses and seminars is highly variable in quantity, subject matter, and geographic location. Courses are usually generalized for a broad audience and may not be fully applicable to the specific needs of each maintenance organization. When considering the use of a seminar for large groups of maintenance personnel, be sure to screen the proposed course syllabus or send limited attendees to the course before committing for full attendance. Pre-screening a course sometimes results in the development of a customized course that can be taught in-house.

Another source of training is the major manufacturers of various equipment and systems. Courses given by the manufacturer are usually taught at the workplace and are well suited for maintenance personnel whose primary responsibilities involve the specific equipment. Where a formal course does not exist, often training can be arranged through direct contact with the equipment manufacturer.

Summary

Whether a facility is large or small, its maintenance organization must be staffed for all of the direct and indirect maintenance activities and functions. As the size and complexity of the facility increases, the staff grows accordingly. Once an organization is staffed, productivity is maintained or improved through management of the maintenance personnel. Job descriptions aid both employees and management in understanding the responsibilities and duties of each position. Feedback from management regarding individual performance in the form of standard evaluations leads to higher quality work output. Effective training augments employee capabilities and reduces maintenance costs.

Chapter Three
ENGINEERING CONSIDERATIONS

Chapter Three

ENGINEERING CONSIDERATIONS

The majority of facilities maintenance costs are determined by the basic design, construction, and present condition of an existing facility. Thus, the facilities manager has few opportunities during the life of the structure to significantly alter general cost factors in the maintenance program. It is most important, therefore, to take full advantage of these opportunities as they arise. This is the role of engineering in facilities maintenance—*facilities engineering*. Engineering is applied in the areas of new design, modifications, improvements, replacement, and in the troubleshooting phase of repair. The level of engineering effort varies for each of these phases and is often divided between staff engineers and engineering consultants hired for specific tasks. This chapter contains discussions of each of these phases.

Engineering the New Design

The primary consideration of a new design is the production of a functional facility; a new or changing need for the organization has prompted the construction of expanded or additional facilities. The definition of these needs and how these needs must be functionally satisfied is generally well documented and easily communicated to the architects and engineers. Since the functional requirement drives the new design, there is a tendency to concentrate on this aspect of the design at the expense of other considerations. However, the conceptual phase of a new design is the single point in time at which there is total control over the future maintenance and operating costs of the facility. For this reason it is essential that the maintenance manager be a member of the design team, and that maintenance considerations and objectives be well stated at the earliest point possible in the design process. The ability to control or modify maintenance costs diminishes from the earliest step in the design process as noted in Figure 3.1.

Life Cycle Cost of Facilities

For new design projects the life cycle cost consists of four phases, which are listed below.
- initial construction
- operating cost
- future modifications
- maintenance costs

Control Over Long Term Maintenance Costs
During the Life of a Facility

Stage of Facility	Degree and manner of control over long term maintenance costs
CONCEPT	**NEAR TOTAL CONTROL** The type of facility can be tailored to any maintenance philosophy, with any feasible system, equipment, or type of material.
DESIGN	**NEAR TOTAL CONTROL** The limits on the design are only those imposed by the approved concept. This may have set exterior appearance, exposure, building height, etc., but flexibility remains to choose systems, etc. which have low maintenance costs.
CONSTRUCTION	**VERY LIMITED CONTROL** At this point the long term maintenance costs have been engraved by the design. Thorough inspection during the construction phase which will preclude premature failures, leaks, cracks, etc.
OCCUPANCY	**PRACTICALLY NO CONTROL** Only education of the users will help to reduce maintenance costs. Prompt reporting by the users of minor problems in the facility can allow prompt repair before the problem grows.
REPAIR	**VERY LIMITED CONTROL** Minor corrections to maintenance intensive items may ease costs.
REPLACEMENT	**LIMITED CONTROL** During replacement of a single item, a system or piece of equipment which is less costly to maintain can be installed.
IMPROVEMENT	**LIMITED CONTROL** By its nature, an improvement project is performed to reduce operating and maintenance costs.
MODIFICATION	**LIMITED CONTROL** System and equipment options may be limited by the party requesting the modification.

Figure 3.1

The cost of *initial construction* is the largest single expenditure in the life of the facility, however, it is not necessarily the highest overall cost incurred during its life cycle. The *operating cost* includes all labor necessary to keep the facility or equipment in normal operation. It also includes the cost of utilities. The cost of housekeeping and consumables may also be included in this category. The third element is the cost of any *future modification* and improvement of the facility. Although the facility is designed for a specific purpose, consideration must be given during the design stage to the likelihood that the facility will require future modification. Modification might be required to accommodate changing needs or advances in technology. The possibility of modification is often fairly predictable but, if ignored, can lead to unanticipated costs. In those cases where a facility has a predictable life span and no foreseen future use, the cost of demolition and removal must be considered during the design phase. The final element is the *maintenance cost* of the facility. The maintenance cost includes the cost of minor repair, major replacement, and preventive maintenance.

Generally stated, an increased investment at initial construction can yield a dramatic reduction in future maintenance costs. Figure 3.2 provides a life cycle cost comparison between bituminous paving and concrete paving for a sample project. It can be see that in this case the increased initial cost results in a lower total cost over a 30 year period. Such decisions require considerable planning in the design stages, or *designing for maintainability*. Otherwise, the actual long term maintenance costs may evolve as a matter of fact, rather than as a planned event.

Design for Maintainability

A competent design team will, in its layout of the facility and selection of materials, be considering the long term life of the facility. This means addressing both function and aesthetics. The intended use of the facility, the length of its desired useful life, and the local climate guide the architects and engineers to choose an appropriate design. Exterior walls are designed of durable, low maintenance materials. Interior floors are tiled or carpeted based on anticipated traffic and wear, with special attention paid to areas where durability to withstand frequent use is a foremost concern. These are maintenance concerns. However, few architects and engineers working on new construction designs have had extensive experience in designing for maintainability. Once the project has been constructed there are few reasons to provide feedback to the architect about the design when maintenance problems arise. For this reason, it is important to have a maintenance manager on the design team.

In order to strike the proper balance between initial investment and continued maintenance costs in designing for maintainability, the design team should first formally identify the maintenance philosophy, or objectives intended for the facility. The maintenance philosophy may be dictated by elements other than the choice of materials or equipment. For example, the maintenance philosophy for a new addition to an existing facility should be similar to maintenance methodologies already in place for the older facility. The same mix of contract services and in-house staff should be expected to provide the repair and preventive maintenance services. For the organization with little turnover in the maintenance work force, this means that similar types of equipment should be installed in the addition to match the skill level of that staff. It would be unrealistic to impose a pneumatic heating control system in a facility

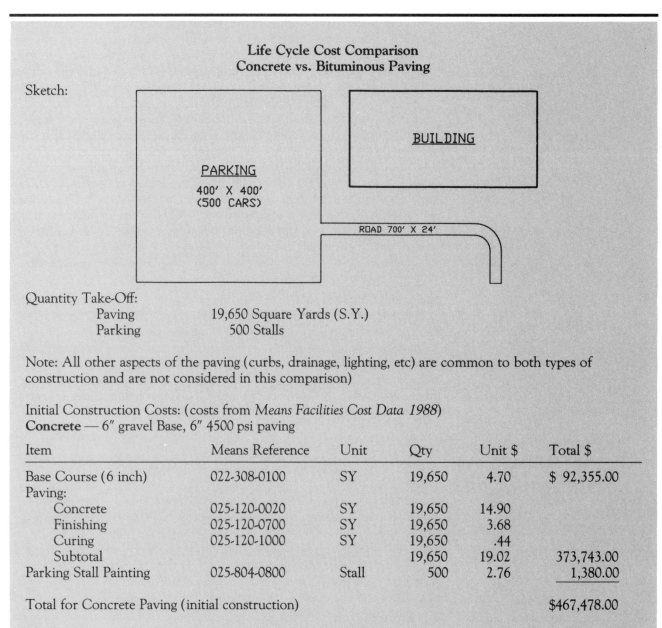

Life Cycle Cost Comparison
Concrete vs. Bituminous Paving

Sketch:

BUILDING

PARKING
400' X 400'
(500 CARS)

ROAD 700' X 24'

Quantity Take-Off:
Paving   19,650 Square Yards (S.Y.)
Parking   500 Stalls

Note: All other aspects of the paving (curbs, drainage, lighting, etc) are common to both types of construction and are not considered in this comparison)

Initial Construction Costs: (costs from *Means Facilities Cost Data 1988*)
Concrete — 6″ gravel Base, 6″ 4500 psi paving

Item	Means Reference	Unit	Qty	Unit $	Total $
Base Course (6 inch)	022-308-0100	SY	19,650	4.70	$ 92,355.00
Paving:					
Concrete	025-120-0020	SY	19,650	14.90	
Finishing	025-120-0700	SY	19,650	3.68	
Curing	025-120-1000	SY	19,650	.44	
Subtotal			19,650	19.02	373,743.00
Parking Stall Painting	025-804-0800	Stall	500	2.76	1,380.00

Total for Concrete Paving (initial construction) $467,478.00

Bituminous — 9″ Base Gravel, 4″ Base Asphalt, 2″ Wearing Course

Item	Means Reference	Unit	Qty	Unit $	Total $
Base Course (9 inch)	022-308-0200	SY	19,650	7.00	$137,550.00
Paving:					
4″ Base Course	025-104-0200	SY	19,650	7.20	
2″ Wearing Course	025-104-0380	SY	19,650	4.07	
Subtotal			19,650	11.27	221,455.50
Parking Stall Painting	025-804-0800	Stall	500	2.76	1,380.00

Total for Bituminous Paving (initial construction) $360,385.50

Figure 3.2

Life Cycle Cost Comparison (continued)

Maintenance Requirements:

Concrete:
    Joint Sealing (rake joints & seal) — every 10 years
    Paint parking stalls — every 3 years

Bituminous:
    Seal Coating (2 coats of coal-tar emulsion — every 3 years)
    1" wearing course overlay — every 15 years

Life Cycle Comparison — 30 year life assumed

Concrete:
Initial Construction	$467,478.00
Joint Sealing at year 10	10,000.00
Joint Sealing at year 20	10,000.00
Paint stalls — 10 times	13,800.00
Total	$501,278.00

Bituminous:
Initial construction	$360,385.50
Seal Coating — 10 times	98,250.00
Overlay at year 15	44,409.00
Painting stalls	13,800.00
Total	$516,844.50

Summary:

Despite a greater initial construction cost, concrete paving is over $15,000 less costly than the equivalent bituminous paving system over a thirty year life. This comparison fits only this example and similar calculations should be made for individual applications.

In addition to the lesser costs of the concrete system, there will be fewer occasions during which the lot will be out of use due to maintenance, a major consideration if the lot is used by workers year round.

Figure 3.2 (continued)

where all existing controls are electrical. The need for a completely new inventory of spare parts, coupled with the need to re-train the work force could be very expensive. These costs must be compared to any marginal decrease another system may offer in initial costs. Thus, at the conceptual stage, the design team should integrate existing and predicted maintenance techniques into the design. The team should consider the personnel, methods, and equipment used by the organization before making any important design decisions.

Standardization of components and systems within a facility has been shown to reduce maintenance costs by reducing the required spare parts inventory. It also allows workers to gain greater proficiency in the repair of a single type of equipment. In order to ensure that any additional facility will have similar equipment, the maintenance manager should maintain an updated set of as-built drawings and manuals containing the listing of preferred standard equipment. This listing might be organized according to the MASTERFORMAT of the Construction Specification Institute, Inc. These classifications are listed below.

CSI MASTERFORMAT DIVISIONS

Division 1 — General Requirements
Division 2 — Site Work
Division 3 — Concrete
Division 4 — Masonry
Division 5 — Metals
Division 6 — Wood & Plastics
Division 7 — Moisture-Thermal Control
Division 8 — Doors, Windows, & Glass
Division 9 — Finishes
Division 10 — Specialties
Division 11 — Equipment
Division 12 — Furnishings
Division 13 — Special Construction
Division 14 — Conveying Systems
Division 15 — Mechanical
Division 16 — Electrical

The listing should include the type of equipment or material, its manufacturer, the model or series identification, and the commercial standard describing the product. Catalog cuts of previously furnished equipment and materials are particularly helpful since these are less susceptible to misinterpretation. When equipment is organized in this manner, it can be easily identified. This will enable current standards to be incorporated into future designs, ensuring continuation of standardized equipment, fixtures, and materials.

This composite listing should be used when reviewing new designs and should be conveyed to any architectural or engineering firms on retainer for new work.

Selecting the Architect/Engineer

The decision to hire an architectural or engineering firm to perform a design often arises from a lack of in-house capability. The facility manager may have no formal design staff, or the staff may be too busy, too small, or too inexperienced.

The facility manager should realize that hiring an outside firm to perform the design may present several challenges. Unless only routine designs are required, the outside firm may not be able to initially and completely understand the intended use and function of the desired facility. The architect/engineer, therefore, should study existing procedures. This may entail extensive interviews with the facility's staff. The schedule for the design, a reasonable budget for construction, and the relationships between the facility manager (the owner), the design team (the architect/engineer), and the contractor constructing the project should be clearly defined. In addition, the architect's level of involvement during the construction phase and the amount of payment for the design should also be clearly stated before construction begins.

Formal accredited educational programs and registration requirements provide established levels of professionalism within the architectural engineering community; thus, most architectural engineering firms are capable of providing a design to meet the client's needs, and will have a series of completed projects which point to their competence. So how does a facility manager select the right firm to employ for a particular new construction or major modification project? Publicly funded organizations must abide by prescribed procedures to ensure fair competition between interested architectural or engineering firms. The Federal Acquisition Regulations, for example, provide firm guidelines for federally funded design projects. Regardless of whether or not the procedures are formally defined, numerous factors should be systematically considered.

Getting Started

If the organization has already used a particular engineering firm, was satisfied with the product that firm produced, and is not bound by legislative restrictions regarding open competition for additional design work, the original firm would be the logical point to start. Should that firm be unavailable or unqualified for the particular project, however, a search must be made to identify a group of eligible firms.

Typical initial sources for candidate firms include the yellow pages, consultation with other facility managers, or placing a request for proposals in local classified ads. For projects which deal with a particular specialty of engineering it may be necessary to extend the search beyond the local area. Advertisements placed in national trade publications often generate substantial interest in the project.

When contacting potential architectural or engineering firms, the scope of services or statement of work requested should be clear. The statement of work allows an interested firm to determine whether it has the capability to perform the design and whether to pursue the job. This document also provides a basis for an eventual contract between the facility owner or manager and the design firm. Figure 3.3 is an outline of a typical statement of work. This statement is often subject to modifications and used for negotiations during and after the selection process.

Narrowing the Field

Upon initial contact, potential architectural or engineering firms should be given the basic statement of work (outlined in Figure 3.3) and should be asked to submit an information packet describing the firms' personnel, qualifications, specialties (if any), and past work experience. There are specific forms for federally funded work (Standard Forms 254 and 255), which organize this information in an easily comparable format. More

OUTLINE STATEMENT OF WORK FOR
ARCHITECT-ENGINEER SERVICES

I. SCOPE OF WORK
    A. Services Desired
            Planning Study
            Conceptual Design
            Complete Design
    B. Type of Work
            New Construction
            Renovation
            Addition
    C. Size of Project
            By Square Foot
            By Functional Parameter
                Number of Work Stations
                Size of Plant (3 megawatt, 400 horsepower, etc.)
    D. Location of Project
            City, State
            Street Address (if significant)
            Company Name
    E. Contracting or Inspection Services Required
    F. Construction Budgets
            Funds Available
            Redesign Requirements

II. Time Schedule - Submission Requirements
            Architect-Engineer Contract Award
            Design Submittals
                Conceptual
                    Number
                    Extent
                    Type of Cost Estimate
                50% Submittal
                90% Submittal
                100% Submittal
            Owner Review of Design
            Solicit Bids
            Open Bids
            Award Construction Contract
            Commence Construction
            Occupy

III. Method of Contracting
            Competitively Bid
                Open Competition
                Limited, Prequalified Bidders
                Small Business Only
            Pre-selected Contractor
                Negotiated Fixed Price
                Cost Plus a Fee
            Constructed by Owner

IV. Specifications and Drawing Requirements
    A. Specifications
            Format Uniform Construction Index, Construction
                Specification Institute, etc.
            "Generic" No Brand Names Allowed
            At Least Three Equal Products

Figure 3.3

```
                    Performance Specifications
                    Referenced Commercial Standards
                    Proprietary Items Allowed?
        B.  Drawings
                    "D" Size Originals
                    Quantity for Each Design Submission
                    Quantity for Bidders
                    Type of Paper, Vellum, Mylar, etc.
                    Completed, Reviewed by Engineer/Architect Regis-
                        tered in State of Construction

    V.  Additional Required Services
        A.  Local Permit Applications
                    Planning Board
                    Zoning Board
                    Prepare Documents
                    Represent Owner at Meetings
        B.  During Construction
                    Review of Contractor's Submittals
                        Schedule of Values - Cost Breakdown
                        Samples/Materials/Certificates of Compliance
                        Shop Drawings
                    Site Visits
                        Period Inspection
                        Final Inspection
                        Punch List Preparations
                    Consultation During Construction
                        Review of Change Orders/Changed Condition
                        Review of Value Engineering Proposals

   VI.  Cost Estimates
        A.  Life Cycle Cost Estimates
                    For Major Design Alternatives
                    For Reduction of Maintenance Costs
        B.  Design Estimates
                    System Estimates at Conceptual, 50% Design
                    Detailed Unit Price Estimates at 90% and Final
                    Breakdown by Specification Section
                    Breakdown by Probable Subcontract Area

  VII.  Payment Schedule
        A.  Lump Sum for Design to Include:
                    Design Preparation
                    All Submittals
                    Preparation of Permit Applications
                    Representation at Planning Boards, etc.
                    Preparation of Cost Estimates
                    All Site Visits for Design Preparation
                    Designated Number of Construction Inspection Visits
                    Correction of Drawings to As-Built Condition
                    Submittal Review During Construction
        B.  Milestone Payment Schedule and Percent Paid
                    Completion of Planning Board, Permits, etc.
                    Conceptual Design
                    Intermediate Design Submissions
                    Final Design Submissions
                    Intermediate Design Submission
                    Intermediate Payment During Construction
        C.  Payments for Additional Services
                    Hourly Rates, Overhead Rate
                    For Additional Design Reviews
```

Figure 3.3 (*continued*)

firms have these forms already prepared. Firms which do not have such forms available usually have a package describing the firm's qualifications and past projects. The information package should contain the following information:

- Number of persons employed by the firm.
- Qualifications of the principals and primary architects and engineers. For a firm of fewer than 15 employees, the specialty and qualifications of all professionals are generally provided. For larger firms this information may be summarized by category of architects, civil engineers, mechanical engineers, structural engineers, etc.
- Age of the firm.
- Location of the firm and branch offices (if any).
- A summary of all projects completed within the past three years, regardless of type. This summary should include a brief description of the project, square footage, construction cost, date of design, date of construction, and the firm or person who commissioned the project.
- A summary of all projects completed which resemble the project in question.

Information packages from several firms give the facility manager enough information to begin a formal comparison of the firms. For federally funded work, the total number of interested parties must be narrowed down to a *short list* — 4 to 10 firms that receive further consideration. This process is helpful for all projects because it results in a more manageable number of highly qualified competitors. Some suggested criteria on which to base this first selection are listed below:

- **Ability:** Is the firm able to perform the project? A firm which usually performs small design work may not be capable of making a major step upward in project scale. Similarly, a firm which has worked mainly on hospital design may not be the best choice for the design of a new elementary school.
- **Level of Expertise:** Do all of the firm's professionals possess the proper level of expertise? Have they obtained a professional registration and license to practice architecture and/or engineering in the local area?
- **Location:** Where is the firm located? In the course of preparing a design, the architects and engineers should meet frequently with the end users of the facility. The location of these meetings usually alternates between the project site and the architect's offices. During the design phase several site visits are required. Once construction begins, the architect should make frequent site visits to ensure compliance with the design and to resolve any discrepancies. For these reasons, it is generally most convenient to use a local firm.
- **Number of Employees:** The number of persons within the firm is a simple indication of the firm's capabilities, but bigger is not necessarily better. A large firm can present impressive lists of highly qualified individuals, but only a few of those individuals actually work on each project. Since large staffs indicate large volumes of work, it is less likely that the principals of the firm will be intimately involved in your project design. A small firm, on the other hand, may not have a qualified person to perform the design. This becomes obvious when the design staff is reviewed. If the firm lacks a particular engineering discipline

within its staff, it will have to augment with an outside firm or individual in order to perform the job. Such augmentation is not necessarily a problem, however, since the architectural or engineering firm usually seeks a specialist in the discipline needed. A disadvantage of this situation is that the design team may become somewhat dispersed and the overall integration of the design may suffer.

At this point in the selection process, the facility manager should form a selection committee. The committee should include one or more representatives of the facility's users. For an addition to an elementary school, for example, this representative might be the school principal or a senior teacher. The facility maintenance manager should also be included. The facility manager or facility engineer should chair the committee. It may also be desirable to have a member of the facility's purchasing department present.

The committee should meet prior to receipt of the initial proposals from potential firms, to establish selection criteria for the short list. Some standardized form of scoring (or simple high-medium-low sorting) each firm based on these criteria should be agreed upon. Each member of the committee should then evaluate each of the potential firms based on the established criteria. Thereafter, the committee should collectively review each submission. In rare cases, the results of the preliminary review are so distinctly in one firm's favor that further competition is unnecessary. Normally, however, this first stage in the selection process identifies several well qualified firms.

Interviewing the Finalists

Up to this point in the selection process there has been little personal contact with any of the firms. The evaluation which earned each architectural or engineering firm a place as a finalist was based on written communications and presentations. But written presentations can be misleading. It may not be apparent, for example, that the 15 person firm presented on the initial submittal is, in fact, a joint venture of several independent architects and engineers. For this reason it is suggested that a formal meeting with the finalists take place at the firms' offices. At least two hours should be allowed to conduct each individual interview. Some typical interview questions are listed in Figure 3.4.

Further Research

At this point almost all of the information concerning the firms under consideration has come directly from those firms. Unless the facility manager has recent experience with one or more of the firms, some additional research is needed. This research can be accomplished in several ways. The selection committee should visit several of each firm's recent projects. Whenever possible, the building users and owner should be contacted to determine their level of satisfaction with the facility. If possible, the maintenance staff should be asked to submit their opinions. Finally, the construction contractor should be contacted to find out if any unusual problems were encountered during construction due to design deficiencies. Although this information can be extremely useful, it is often difficult to obtain. If problems are uncovered during meetings with the users or owners of other projects, the architectural engineering firm should be allowed to explain.

Suggested Topic Areas for Interviews with Prospective Architectural Engineering Firms

Scenario: Although this is an interview, the A/E firm will prepare an extensive briefing, or sales pitch. This is not to be discouraged. Listen carefully for answers to the below questions, and at the completion of the presentation, ask the questions below which remain unanswered.

Personnel
1. Meet key personnel to be assigned to the project.
2. Determine team leader, who will be key point of contact between A/E firm and owner.
3. Determine the types of projects the team members have recently completed.
4. Determine the extent of experience for the team leader. Has this individual lead a multi-disciplined design team before or is this the first such venture?
5. Verify professional registration of the design team members.
6. Has any member of the design team had significant experience in the maintenance of facilities?

Design Office Facilities
1. Ask about the mechanics of the design preparation.
2. Will all drafting be done in-house?
3. Does the firm use a Computer Aided Design and Drafting system?
4. Does the firm use standard specifications? If so, which format? (The choice is up to the owner, not the A/E firm, so this question will help determine if the firm can produce a suitable set of specifications which meet the owners needs.)

Design Procedures
The statement of work for the project which lead to the interview will have stated the schedule and content of design submittals. This is the opportunity for the A/E firm to explain how they plan to meet the various deadlines.

1. Talk about turn-around time for submittals. Can the A/E firm provide rapid approval/disapproval of submittals?
2. Verify intent for attentance at progress meetings during construction.
3. Determine if the A/E firm has other projects which will compete with this project for time and attention.

General Details
The design team and the owner will work closely together for the term of the design and subsequent construction. The interview is the time to determine if the two parties are compatible.

Figure 3.4

Final Selection

The final selection is based on the conclusions drawn by the committee. The committee should meet first for a general discussion of their impressions of the finalists' capabilities. This discussion should be frank and open. The objective is to determine the strengths and weaknesses of each firm under consideration. The discussion may be recorded for future reference.

Following this discussion each member should rank the finalists. The committee chairman should determine if a consensus opinion can be reached. Continued discussion may be necessary before a final ranking is established. Phone calls to some of the firms may be necessary to clarify any last minute questions. The final selection is often a compromise between two or more firms with individual strengths and weaknesses. By referring back to notes taken during the open discussion about the

Suggested Topic Areas for Interviews with Prospective Architectural Engineering Firms (continued)

1. Express your maintenance philosophy. Is the A/E firm prepared to follow that philosophy? If not, find another firm.
2. Does the firm have experience with the design of facilities using the systems and fixtures which you prefer or demand?
3. Ask for life cycle cost estimates, including all future maintenance costs. Ask to see other such estimates prepared by the firm for other projects. If none are available, this is new ground for the firm.
4. Verify once more the work location of all team members. if the team is a joint venture from geographically separated offices, determine how they will meet to integrate design details.
5. For rehab work, determine how many site visits the team will make during design preparation. Determine how they will clear up questions on dimensions as design proceeds. (Trips away from the design office are expensive to A/E firms since there is little productive output from the time spent. Therefore the A/E firm will desire to minimize such trips. During rehab projects on existing conditions, some will be missed during initial visits. Follow-up visits must be made to clarify unknowns. If skipped, costly changes can result during construction.)

Closing the interview

Regardless of your satisfaction or dissatisfaction with the interview results, do not make or convey a decision to the firm at the time of the interview. If pressed, politely advise of your intended date for making a decision.

Figure 3.4 (continued)

finalists, the committee may identify particular precautions to be taken with the top finalist. Although the best firm has been identified, the entire list should be rank ordered, in the event that negotiations with the chosen firm prove unsuccessful. In this case, similar negotiations would begin with lesser ranked firms until an agreement can be reached.

The Design Fee

Once the architectural or engineering firm has been chosen, the process of negotiations can begin. The fee for design services, the schedule of completion, the staff members assigned to the project, and the specific roles of the design firm, the owner, and the contractors are important negotiating concerns. The method of competitive bidding is often desirable to procure the best product at the lowest price. Competitive bidding, however, for design services is very rare and for good reason. The facility manager or owner does not wish to encourage competitive bidding, for fear that the engineering firm, in an effort to provide a design within a competitive low bid, would be forced to take short cuts which might jeopardize the integrity of the building and the safety of its occupants. The fee, therefore, is a function of the size and complexity of the project desired. The fee proportionately increases as the size of the project increases. Similarly, the fees increases as the architectural and engineering complexity increases. For example, the design fee for a $2 million hospital project will be considerably higher than the fee for a $2 million warehouse. Figure 3.5 is extracted from *Means Facilities Cost Data, 1988* and shows typical architectural and engineering design fees, expressed as a percentage of total construction costs by size and type of construction. These figures are mid-range fees for negotiating the final costs of these services.

Schedule for Completion

The design professional and the owner must agree upon a reasonable schedule for completion of the design and, hence, commencement of construction. The schedule should not be negotiated, however, but merely confirmed at this time. The schedule for design submittals should have been included in the initial statement of work given to the design professional during the selection process (outlined in Figure 3.3). The design firm, therefore, has accepted the job aware of the required time constraints. Under normal circumstances, the design firm should be capable of completing the design within the time limits set by the facility manager. However, since the format for coordinating user needs with the design firm can vary, it is possible that agreement on a design submittal schedule could be difficult. The schedule, therefore, should always be a requirement of the design, not a point for future negotiation.

The Design Staff

An additional point which bears clarification at final negotiations concerns the design staff. Both the owner and the design firm must agree upon those individuals from each party who will be participating in the effort to produce a workable design. During the initial submittal and further during the interview, the design firm presents the most capable members of its staff. It is essential that, after selection, the participation of these key persons, who may have influenced the selection of the firm, continues through the design and construction effort.

The Role of the Design Firm

A final point to be clarified during negotiations is the role of the design firm during construction. After the design is completed, the architectural or engineering firm is required to review the various submittals and changes requested by the construction contractor during construction. The contractor is required by the construction contract to submit material samples, manufacturer's data, and various certificates of compliance. Normally these submittals are reviewed and approved by the architect or engineer. Occasionally, a dispute requires mediation by the owner. This process assures the owner that the materials and methods used are in accordance with the contract requirements. Thus, specific procedures for handling the submittals and resolving potential disputes must be negotiated and written in the contract before construction begins.

Another responsibility of the design firm is to review contractor requests for regular progress payments. The contractor presents an invoice, usually monthly, for payment for all work completed and all materials delivered to the site up to the date of the invoice. The design firm reviews that submittal to verify that the work has been completed and that the amounts shown on the invoice represent fair value for the work done.

Final Instructions to the Architect/Engineer

Once the architectural or engineering firm has been selected and the design fee and statement of work negotiated, it may be helpful to restate the basic premises for the design to the design team. When the design team starts with a clean sheet of paper, they may be aware of the thousands of options available and may be tempted to create an innovative design. This is commendable, but only if the resultant design can be constructed at a cost similar to that of conventional construction

---

⑩ Architectural Fees (Div. 010-004)

Tabulated below are typical percentage fees, below which adequate service cannot be expected. Fees may vary from those listed due to economic conditions.

Rates can be interpolated horizontally and vertically. Various portions of the same project requiring different rates should be adjusted proportionately. For alterations, add 50% to the fee for the first $500,000 of project cost and add 25% to the fee for project cost over $500,000.

Architectural fees tabulated below include Engineering Fees.

Building Type	Total Project Size in Thousands of Dollars						
	100	250	500	1,000	2,500	5,000	10,000
Factories, garages, warehouses repetitive housing	9.0%	8.0%	7.0%	6.2%	5.6%	5.3%	4.9%
Apartments, banks, schools, libraries, offices, municipal buildings	11.7	10.8	8.5	7.3	6.7	6.4	6.0
Churches, hospitals, homes, laboratories, museums, research	14.0	12.8	11.9	10.9	9.5	8.5	7.8
Memorials, monumental work, decorative furnishings	—	16.0	14.5	13.1	11.3	10.0	9.0

Figure 3.5

while providing for the intended function. The designers should, therefore, be cautioned that the owner's primary concern is to obtain the lowest possible life cycle cost for the facility. The architects and engineers should be reminded that the future maintenance of the facility must be considered at every step of the design process.

Construction Management

The procedures previously outlined have implied that the facility manager has sufficient staff with enough expertise to effectively manage the selection of the architectural or engineering firm and to administer the design and construction contracts. For many organizations, however, the need for this expertise is insufficient to justify hiring and maintaining a facility maintenance staff on a full-time basis. A hospital administrator, for example, would probably face a major construction effort only two or three times in the life of the building. In between these major construction efforts, there may be few instances where engineering design expertise is required. Thus, the field of construction management has evolved to protect the interests of inexperienced owners when administering design and construction contract. The construction manager, or construction management firm, is hired to perform the following tasks:

- Coordinate relations between the owner and the architect in the development of design specifications and drawings.
- Review contract specifications and drawings for compliance with owner's needs.
- Review the drawings for "constructability," identifying any changes in design which would reduce the cost of construction while satisfying the owner's needs.
- Develop a list of potential bidders and invite bids for the construction. The construction manager may also decide to subdivide the project into separate contracts to reduce costs. The construction manager is then responsible for coordinating these separate contractors working on the same site.
- Review all of the contractor's submittals, particularly the schedule of values, which breaks down the contractor's total bid into costs for each individual work area.
- Review the contractor's project completion schedule. Keep track of the contractor's compliance with that schedule, issuing proper notices when the contractor falls behind the planned completion schedule.
- Regularly chair and conduct job progress meetings between the contractors, architects, engineers, and the owner.
- Maintain daily records of work accomplished, personnel present, material delivery, and climate conditions.
- Inspect all work for compliance with contractual requirements. This would include inspection visits to off-site fabrication areas or shops.
- Maintain cumulative records of all work completed to date. Certify contractor's submittals for payment. This includes verifying the amount of work claimed to have been accomplished and the dollar value assigned to that work by the contractor.
- If authority to expend funds has been given to the construction manager, he will make all payments to the contractor subject to the review noted above.

- Note any errors and omissions in either design or construction and recommend or order the solutions to such errors.
- Mediate and resolve any disputes between or among the architect, engineer, contractor, or owner.
- Ensure that required permits are obtained by the contractors prior to construction.
- Make recommendations to both the architect and contractor concerning methods to reduce cost or hasten the completion date of the project.
- Oversee safety procedures and precautions on the job site.
- Review and rule on any change orders necessary to complete the work.

The method for selecting a firm or individual to perform construction management is similar to the procedure outlined for procuring A/E services. The emphasis should be placed on the candidate's past record of handling similar work. Fees for construction management services vary from approximately 10–15 percent of construction cost for small projects under $10,000 to 2–5 percent for projects costing $1,000,000 and up.

Engineering Modifications and Improvements

A modification requested by a facility user is an opportunity for the facility maintenance manager to incorporate components into the building which can reduce its life cycle cost. If, for example, a modification involved re-configuration of some office spaces, new walls could be constructed of more durable but easily removable partitions. This might reduce future painting costs and allow for easier re-configuration of the space.

A second major engineering factor to be considered when making modifications to an existing facility is the standardization of components and equipment discussed in Chapter 1. If the component in the modification is not currently standardized, the facility manager should take this opportunity to initiate a standardization program.

Engineering is the essence of improvement projects. Improvement projects, as stated in Chapter 1, are initiated by the facility maintenance manager or engineer with the objective of reducing the long term operating and maintenance costs. Improvement projects result from an engineering analysis and comparison of alternative systems.

A common impetus for improvement projects is the incorporation of more energy efficient components or systems. These systems and components are often advertised in architectural and engineering trade journals and magazines. The engineer should examine the claims and potential for each new product advertised. Only after a favorable life cycle cost analysis is complete should the components be considered. When a facility manager does not have a staff engineer, a consultant can be hired to examine a particular segment of the facility operation and facility maintenance program to determine possible improvement projects. This form of consulting is very common in the field of energy analysis. Energy efficiency improvements generally pay for themselves in a relatively short period of time.

Engineering and Troubleshooting

A majority of the maintenance problems within a facility can be solved by the craftsman who maintains its components or systems. An equipment failure can often be attributed to improper installation, improper maintenance, or the simple fact that the equipment has reached the end of its normal useful life. In addition to repairing one such piece of equipment, the facility maintenance manager should check for potential problems in similar components or installations throughout the facility.

When a component has a continual problem which has not been resolved by the maintenance staff, it may be necessary to hire an engineer. The engineer will survey the problem and perform calculations necessary to determine the actual cause of the failure. An engineer's analysis may lead to the detection of an unknown problem, such as an untraceable leak in a built-up roofing system. Such analyses are easily performed, but often expensive, diagnostic equipment is required. This equipment is usually owned by a consultant who specializes in the particular field of the problem. Despite the initial expense, the use of a specialist can save a considerable amount of money and time when hired early to solve a problem.

Engineering History of a Building

A major engineering casualty or failure is the fastest way, but also the worst, to learn about the engineering layout and history of a facility. When a pipe bursts, there is little time to find the dusty set of original construction drawings and start tracing piping systems. If such drawings are not available, the problem is even harder to solve. Thus, it is desirable, even essential, to maintain a proper record of the engineering history of a facility and its systems and components before anything goes wrong. A complete facility history consists first of the original as-built construction drawings. Any modifications made to the facility should be noted on these drawings. This engineering record should also include the history of equipment components, including the date of initial installation and any overhaul, replacement, and preventive maintenance performed on the equipment. The last portion of the equipment history is often the best indicator of why a component has failed. Unfortunately, some failures occur shortly after a preventive maintenance visit, due to faulty workmanship. This should not be considered an indictment of the preventive maintenance program. It must be accepted as a normal result of a vigorous project management program. The facilities manager must realize that equipment which is properly maintained should last to and beyond it's predicted useful life.

Inspections

A final method of engineering troubleshooting is the performance of regular inspections. A non-engineer can recognize many of the problem areas and, with a properly prepared checklist, can perform most of the required inspections. For structural inspections, however, a licensed structural engineer should be employed. Such a program of regular inspections can prevent countless problems. Therefore, it is helpful to establish a formal inspection program. Records of the inspections should be retained for future reference.

Summary

Whether a facility has its own engineering staff or occasionally hires engineering services, the role of architects and engineers in providing the most economically maintained facility cannot be overstated. The key to successful employment of engineering talent is to select the engineer carefully and provide specific task instructions regarding the scope of services required.

Chapter Four

COST ESTIMATING AND BUDGETING

Chapter Four

COST ESTIMATING AND BUDGETING

All maintenance activities require expenditures for personnel, materials, tools, or equipment. Each of these maintenance items has a dollar value for a given year. If there were unlimited resources available to the maintenance manager, there would be no need to consider the cost of each maintenance item. The manager would simply spend the required amounts, when necessary. However, in reality, there are finite limits to the availability of maintenance resources. Applying these limited resources for the greatest benefit is a primary function of the maintenance manager.

The cost of all activities must be accurately understood and predicted as far in advance as possible in order maintain reasonable control over maintenance costs. The manager is then able to make reasonably accurate decisions regarding the desirability of performing any particular task. The cost estimate becomes one yardstick by which the satisfactory completion of the task is measured.

The total amount of all estimated maintenance costs for a particular period comprise the *maintenance budget*. This chapter contains descriptions of the various elements that make up the maintenance budget and explanations of the variables which directly affect those costs. Various methods for accurately estimating the cost of individual activities are discussed in this chapter.

Elements of Maintenance Costs

The overall cost of any maintenance program consists of two distinct cost types. First is the *out-of-pocket* cost of performing maintenance activities. These costs accrue on a daily basis as salaries are paid, materials consumed, tools acquired, and maintenance support spaces utilized. These costs are directly measurable and predictable. The second cost type is the cost associated with the *failure to perform* individual or collective maintenance activities. These kinds of costs are more difficult to directly measure or predict. The cost of a deficiency in maintenance may be the shortened life of a facility component or system. This cost, while predictable, may not be realized for many years. The loss of production or profit due to poor maintenance or slow repair is more easily measured.

Labor Costs

Each employee within the maintenance organization is compensated in some manner. Some personnel, such as those filling managerial level positions, are salaried at a fixed rate for weekly, monthly, or yearly periods. Others receive compensation based on an hourly wage. For all methods of compensation, the cost to the maintenance organization to retain employees significantly exceeds the actual dollars paid directly to the employee. This is realized in the form of various fringe benefits, taxes, and insurance costs which form a "burden" on top of the wages or salaries.

Wages and Salaries

The defined wage or salary of an individual is established by supply and demand. The wages for an individual maintenance mechanic, for example, are generally the minimum dollar hourly wage. This wage will attract and retain personnel with the requisite skills for the job. Wages for these front line maintenance personnel are determined by the market place. The "accepted" wage level is determined by the availability of suitably skilled workers. Since the quantity of workers within this category is relatively high, the need to change the wage scale for these personnel is generally infrequent. Wages vary only when there is either a significant change in demand or when the cost of living changes. An employer paying a low wage relative to local maintenance organizations experiences greater instability in his work force. Unfortunately, there are no immediate indicators as to when one employer is paying a higher wage than necessary.

The salary paid to a *maintenance manager* is, similarly, the minimum value necessary to attract a person with the appropriate background. While personnel receiving wages are more likely to be viewed as fixed costs which produce equal benefits, the salary offered to the maintenance manager is tied to the expected benefits which the prospective manager brings to the organization. A manager with a recognized track record is offered a higher salary based on the anticipated savings his expertise will bring to the workplace. For this reason, salaries may be negotiable. Once established, wages and salaries are very stable for the purposes of preparing cost estimates. For long term budgeting considerations, however, the likelihood of wage or salary increases should be considered.

Social Security

Both the employer and employee contribute to the federal social security system. The employee's share is deducted from his wages or salary, and the employer contributes a matching amount. The funds collected are used to sustain a large federal program of entitlements, including retirement and medical care benefits. There are two components which make up the social security deductions. The basic deduction is based on a fixed percentage of each employee's pay. The second component is an upper limit on contributions expressed as the maximum amount of wages or salary which is subject to social security deductions. For 1987, the deduction was fixed at 7.85% on the first $45,000 of wages each year. The employer matches the exact amount of this deduction on an employee-by-employee basis. Both the deduction rate and the upper limit are fixed by law for any given year. The rates are usually set one or more years in advance; therefore, this additional cost is easily determined.

Worker's Compensation

Worker's Compensation insurance is established to provide employees with protection in case of job-related injury, illness, or death. Worker's Compensation programs are operated by the state through private insurance companies or state-retained insurance funds. The premium for Worker's Compensation is established based on the general risks associated with a particular trade and the safety record of a particular employer. The exact rates are established by the individual state based on historical data of past injury, illness, and death.

Unemployment Insurance

Unemployment programs are state-operated. Their function is to provide sustenance payments to individuals who are laid off a job due to lack of need by their employer. Personnel who are fired are not eligible for unemployment payments. Unemployment premiums are paid to both the federal and state governments. The federal portion, .7% in 1987, is used to sustain a major fund to back up state funded programs. The state premium is based on the prevailing unemployment levels within the state for various classes of employees. The premium also varies according to an employer's record regarding lay-offs.

Medical Insurance

Many employers provide direct payment to a medical insurance company on behalf of the individual employees. This payment is usually made above and beyond the direct wages and not as a deduction from those wages. The costs of medical insurance benefit programs vary widely based on the scope of medical coverage provided. These medical programs can be used as significant leverage against wage increases as they are usually the most highly valued benefit of an employee.

Retirement

Retirement programs also vary widely. Most are based primarily on direct deductions from an employee's wages with a partial or full matching contribution from the employer. When the employee is a member of a labor union, the contributions are made by the employer directly into the union retirement fund.

Employee Absences

Most employees are provided vacation time as a part of their compensation. Often holidays are given to most, if not all, employees equally. When employees are required to work on these holidays, additional compensation is provided in the form of either overtime wages or deferred time off. The amount of non-holiday vacation time earned is often proportional to the longevity of employment.

The cost of vacation time must be considered when developing a budget. Whether the employee is paid a salary or hourly wage, the vacation time represents a period where the employee receives pay, but is not present for work. If an employee earns four hours vacation time for every 80 hours of work, this means that five percent of the work force is absent, but still being paid, at any time. The impact of this absence on maintenance costs is that each hour of maintenance effort *actually* costs the employer five percent more than it would without vacation benefits. Therefore, in preparing a budget, the total hours required for all maintenance work is 5% less than the actual needed hours of employment.

When employees are unionized and the company has no formal vacation pay plan, the union agreement may require that a portion of the employees' pay be deposited by the employer into a designated fund. That employee may then draw upon this fund when vacation is taken. Alternately, the employer may be required by the agreement to pay this sum over and above direct wages. This form of vacation is often termed *deferred wages*.

In a similar category to vacation pay are allowances for pay during times of illness. Formal pay systems such as those of federal and state governments often allow for the accumulation of *sick leave* to be taken as needed. While the cost of providing this time off is not immediately evident, the net effect is that the number of hours actually worked is less than the total hours paid.

Travel

In many instances, extensive travel distances to various work sites exceed those normally associated with home to work travel to a single place of employment. In such cases, special travel pay is provided to the maintenance employee. *Travel pay* is usually not a factor in facilities maintenance work forces. Reimbursement for travel is generally made only in cases where maintenance personnel are dispatched to remote sites. The employee is paid a specified rate for the actual miles driven.

Shift Differential

Shift differential, or *shift premium pay*, is an additional cost paid when work is conducted outside of normal daytime work hours. This premium pay is additional compensation paid to an employee who is working an undesirable shift. The premium is provided either through the direct addition of a small amount to the basic wage or by paying an individual working a late shift for more hours than actually worked. Earning eight hours pay for seven and one half hours worked is a common means of shift compensation.

An ancillary benefit to this means of compensation is that the need for on-site employee parking is reduced. The half hour not worked between shifts allows the parking lot to empty and refill. Parking considerations, however, are usually not a major concern for maintenance organizations. Additionally, some maintenance activities, such as the operation of boilers, treatment plants, or electrical generating equipment require full-time operators and do not allow for a gap between shifts.

Premium Pay Rates

If a certain maintenance task which is performed infrequently requires either increased skills or exposes the employee to increased dangers, a premium is often paid to the employee for these periods of exposure. If, for example, the local prevailing wages for electricians working on high voltage circuits is significantly higher than the wage for low voltage electrical work, the employer might establish that the higher wage is paid only during periods of high voltage work. Although this represents an additional cost, it represents an overall savings since the employer does not have to pay high voltage wages year round.

Overtime

Overtime is paid for hours worked above and beyond the normally scheduled work day or work week. The amount of overtime paid varies from "time and a half," which is a fifty percent increase in wages during overtime, to double or even triple time. Overtime begins to accrue when the employee works more than a certain number of hours per day or a total hours per week. Payment of overtime when an employee works over 40 hours in one week or 10 hours in one day is a typical overtime policy. Overtime is often paid at different rates for different situations. Normal work over 40 hours in one week might be paid at time and a half, while work on holidays is paid at double time.

Miscellaneous Benefits

In addition to the various categories of wage burdens previously listed, numerous benefits provided by the employer increase payroll costs. For example, the cost of providing free parking for employees in an area where commercial parking lots are expensive is not immediately perceived as a personnel cost. The cost for the parking was part of the initial capital investment for purchase of the property and parking lot construction. However, maintenance of the lot is a directly related employee burden. Recreational benefits, employee discounts, and company cars, are also provided as additional compensation to employees. Costs of this type are not usually "charged back" to the portion of the overall organization which derives the benefits. They are generally absorbed as overall organization operating or *overhead costs*.

Materials, Parts, and Supplies

Unlike labor costs, which are directly tied to the number of hours worked on a particular activity, the cost for materials, parts, and supplies varies considerably from one activity to another. These variances are discussed in the following sections concerning individual activity type cost estimating.

Materials

Raw materials are those which are to be formed, molded, cut, fit, or assembled to produce a final product. These materials are purchased in bulk for an individual job or as shelf stock to be drawn upon over an extended period. In general, materials are generic and available from multiple sources. Within this category are common construction materials such as lumber, drywall, nails, concrete masonry units, mortar, and concrete. Pricing for these items is highly competitive, and the materials are usually readily available for immediate delivery from local suppliers in standard type and size.

The consumption of raw materials usually generates some waste, since the materials are provided in standard sizes. When performing minor repair projects, the waste factor may be very high. There might be a large proportion of waste due to the need for odd shapes or sizes to fit the repair, or the standard size of the material may be large compared to the amount actually needed for the application. For example, to patch a small hole in a wall might require the use of as little as 10 percent of a full sheet of gypsum drywall. The amount of waste can be reduced by maintaining a storage space for scraps and cut pieces. The cost of providing this storage is a trade-off to be measured against the benefit of reducing waste.

Parts

Spare parts are maintained on hand for emergency repairs and regularly scheduled replacements. The factor which determines the extent of a spare parts inventory is the availability of the particular spare part measured against the need for restoring the failed component or system to full operation. Establishing a spare parts inventory involves the steps listed below.

1. Identify the components and systems of the entire facility that directly affect the ability of the facility and its users to perform their intended functions.
2. Determine the cost to the facility of prolonged loss of any such component or system.
3. Determine the individual parts of the component or system which are most susceptible to failure, or those parts which, if not replaced, will lead to system failure.
4. Ascertain the ability to purchase the required parts on an as-needed basis. Are parts available in sufficient quantity and can such parts be delivered immediately?

The above analysis will yield a list of the required spare parts to be maintained at the facility.

The price of maintaining a spare parts inventory depends on the source of supply for the individual parts. When the particular spare part is available only from the manufacturer of the system or component in which it is used, the price is stable and often high. When the spare part is available from several manufacturers, the pricing is much more competitive. For example, an expansion valve for an air conditioning system may be available only from the system manufacturer, while belts for the compressor may be purchased from several sources. Some competition in pricing may result from contacting local suppliers for the proprietary parts. The cost of spare parts needed for emergency repairs is often inflated by special shipping or handling charges.

When the need arises to purchase a particular part which was not predicted to fail (and therefore was not in the inventory) the cost of the part is an unavoidable expense. The only decision to be made is whether the cost of expedited shipping is justified in view of the implications of extended down time of the component.

Supplies

Supplies, or consumables, are materials which are regularly used in the operation and maintenance of the facility. Unlike spare parts, consumables are only minimally tied to the components or systems of the facility. This category includes toilet paper, paper towels, cleaning materials and lubricants, which are generally not available from the system manufacturer. The type of flooring material dictates the type of cleaner or wax to be used, but that cleaner or wax is usually available from many different suppliers. The type of paper towel holders installed in the building dictates only the size, not the quality, of the towel to be used. Numerous suppliers provide varied levels of quality and quantities of towels which fit the holder. In developing costs for supplies, the marketplace provides rigorous competition, leaving only the desired quality of the product to cause variations in pricing. A third party organization may establish formal standards for consumable supplies used with certain systems. The Society of Automotive Engineers (SAE), for example, establishes standards for different grades of lubricating oil for automotive engines. These standards are then used by automobile manufacturers to aid car owners in selecting the proper replacement motor oil. In addition, these lubricating standards are used throughout other industries for specifying types of lubricants. The American Society for Testing and Materials (ASTM) establishes standard testing methods and classifications for most types of materials. ASTM standards are used to define the various types of structural steel, for example, used in building construction. The ASTM standard specifies the particular makeup of the steel and the tests and methods of testing which must be satisfied to obtain that rating or classification.

Supplies must be on hand in proper quantities. Quantities are individually determined by need, available storage space, quantity discounts offered at time of purchase, and the ability of the local supply network to respond to the continued need. A summary of the factors which affect the pricing of materials, spare parts, and supplies is shown in Figure 4.1.

Tools and Equipment

Each maintenance activity, from the simple act of replacing a light bulb to the diagnosis of a complex electrical system problem, requires the use of some tool or piece of equipment. Some tools simply make tasks easier and enhance individual worker productivity. Others are essential "tools of the trade", and are unavoidable expenses. Still others offer some enhanced ability to make timely diagnoses of system or component problems. The choice of tools and equipment is made based on the cost of the tool and the benefits derived from its use. Tools represent a

Materials, Parts, and Supplies
Cost Variability

	Source of Supply	Competition Among Suppliers	Quality of Furnished Materials
Materials	Unlimited	Good	Readily specified
Spare Parts	Limited	Limited	Good for exact manufacturer Questionable for others
Supplies	Unlimited	Very strong competition	Variable Predictable where third party standards exist Unknown for others

Figure 4.1

one-time expenditure of funds for which use is derived on many occasions. Simple hand tools are used for many different tasks, while specific testing equipment may be useful for only one specific component.

In general, it is not feasible to directly charge the use of a specific tool to a specific maintenance activity. Only on rare occasions is it economically justifiable to purchase a specific tool for one-time use. Rental of such equipment is the preferred source. A formal method for considering and evaluating the proposed purchase of a specific labor-saving device, tool, or piece of diagnostic equipment is discussed in Chapter 5.

The cost for specific tools and equipment should be included in the maintenance budget. Only when a project is executed on a reimbursable basis is it necessary to determine a pro-rated share of the tool and equipment cost. In such cases, the cost for tools is considered part of the total overhead costs for the project or added directly as single or multiple line items. In practice, the overhead method is preferred for those tools which are a normal part of the shop's inventory. If a tool is purchased for one-time use on a project, the full price is charged to that job. If the tool or diagnostic equipment is rented for the duration of the project, the actual paid rental rate is charged to the customer.

Most organizations establish a *small tools allowance* for each shop to cover the replacement of small hand tools which wear out, are lost, or are stolen. The shop foreman is responsible for the control of the small tools allowance. The maintenance supervisor monitors the consumption of tools and determines if there are indications of excessive loss. Where there is a significant loss factor, a formal small tools inventory control system should be established. For example, in large organizations the control of small tools often requires a tool room with a check out procedure. Employees are required to produce identification, often in the form of an identification card which is surrendered when a tool is issued. The employee is then responsible for the replacement of the issued tool if it is not returned. In smaller organizations, employees are given a lockable tool box for their specific tools. Thereafter, the employee can obtain replacement only for worn tools and is responsible for any losses.

For estimating purposes, the cost of tools can be obtained directly from tool vendors. When it is necessary to charge a pro-rated amount for the use of a specific tool on a specific project, the maintenance organization has several alternatives. The actual hourly, daily, or monthly cost of ownership can be developed through copious formal record keeping. The actual operating costs for the tool can also be calculated. Alternately, the prevailing rental rate and operating costs, as would be charged by a commercial rental firm, can be determined from actual rental quotes or from an accepted tabulation of cost data. Figure 4.2 shows a typical listing of equipment rental rates from *Means Facilities Cost Data, 1988*.

Overhead Costs

As explained in Chapter 2, the maintenance management staff grows as the size of the facility being maintained increases. These supervisory and managerial members are necessary for the proper control of the maintenance work force and maintenance effort, but do not actually directly contribute to maintenance work. The cost of this staff and the support which they require accounts for a significant portion of the maintenance budget and is therefore a specific cost of maintenance in the form of overhead.

016 400 | Equipment Rental

		Description	UNIT	HOURLY OPER. COST.	RENT PER DAY	RENT PER WEEK	RENT PER MONTH	CREW EQUIPMENT COST	
420	6860								420
	6900	Water tank, engine driven discharge, 5000 gallons	Ea.	4.50	190	565	1,700	149	
	7000	10,000 gallons		8.80	320	965	2,900	263.40	
	7020	Transit with tripod	↓		26	77	230	15.40	
	7030								
	7050	Trench box, 8,000 lbs. 8' x 16'	Ea.	1.32	130	395	1,185	89.55	
	7070	12,000 lbs., 10' x 20'		1.63	200	600	1,800	133.05	
	7100	Truck, pickup, ¾ ton, 2 wheel drive		7.75	42	125	380	87	
	7200	4 wheel drive		8.90	57	170	510	105.20	
	7300	Tractor, 4 x 2, 30 ton capacity, 195 H.P.		7.75	290	865	2,600	235	
	7410	250 H.P.		10.70	315	950	2,850	275.60	
	7500	6 x 2, 40 ton capacity, 240 H.P.		12.75	335	1,000	3,025	302	
	7600	6 x 4, 45 ton capacity, 240 H.P.		16.50	420	1,275	3,800	387	
	7700	Welder, electric, 200 amp		.70	17	50	150	15.60	
	7800	300 amp		.80	22	65	195	19.40	
	7900	Gas engine, 200 amp		3.50	32	95	285	47	
	8000	300 amp		3.70	50	150	445	59.60	
	8100	Wheelbarrow, any size			6	19	57	3.80	
	8200	Wrecking ball, 4000 lb.	↓	.30	38	115	345	25.40	
440	4900					—		---	440
460	0010	**LIFTING & HOISTING EQUIPMENT RENTAL**							460
	0100	without operators							
	0120	Aerial lift truck	Ea.	10.25	475	1,425	4,275	367	
	0140	Boom truck	↓	3.65	180	535	1,600	136.20	
	0200	Crane, climbing, 106' jib, 6000 lb. capacity, 410 FPM		19.95	1,025	3,100	9,300	779.60	
	0300	101' jib, 10,250 lb. capacity, 270 FPM	↓	27	1,325	3,950	11,900	1,006	
	0400	Tower, static, 130' high, 106' jib,							
	0500	6200 lb. capacity at 400 FPM	Ea.	43	1,225	3,650	11,000	1,074	
	0600	Crawler, cable, ½ C.Y., 15 tons at 12' radius		13.55	370	1,100	3,325	328.40	
	0700	¾ C.Y., 20 tons at 12' radius		14.05	390	1,175	3,500	347.40	
	0800	1 C.Y., 25 tons at 12' radius		15.10	430	1,300	3,875	380.80	
	0900	1-½ C.Y., 40 tons at 12' radius	⑬	21	670	2,000	6,025	568	
	1000	2 C.Y., 50 tons at 12' radius		23	765	2,300	6,875	644	
	1100	3 C.Y., 75 tons at 12' radius		30	885	2,650	7,975	770	
	1200	100 ton capacity, standard boom		30	1,250	3,725	11,100	985	
	1300	165 ton capacity, standard boom		48	2,000	6,000	18,000	1,584	
	1400	200 ton capacity, 150' boom		90	2,175	6,550	19,700	2,030	
	1500	450' boom		75	2,725	8,175	24,500	2,235	
	1600	Truck mounted, cable operated, 6 x 4, 20 tons at 10' radius		10	545	1,650	4,925	410	
	1700	25 tons at 10' radius		15.20	910	2,725	8,175	666.60	
	1800	8 x 4, 30 tons at 10' radius		20	530	1,600	4,775	480	
	1900	40 tons at 12' radius		20	665	2,000	5,975	560	
	2000	8 x 4, 60 tons at 15' radius		29	725	2,175	6,550	667	
	2050	82 tons at 15' radius		26	1,425	4,275	12,800	1,063	
	2100	90 tons at 15' radius		35	1,150	3,425	10,200	965	
	2200	115 tons at 15' radius		29	1,675	5,050	15,100	1,242	
	2300	150 tons at 18' radius		53	1,275	3,850	11,500	1,194	
	2350	165 tons at 18' radius		33	1,875	5,600	16,800	1,384	
	2400	Truck mounted, hydraulic, 12 ton capacity		16.30	305	915	2,750	313.40	
	2500	25 ton capacity		16.75	510	1,525	4,575	439	
	2550	33 ton capacity		15.05	815	2,450	7,325	610.40	
	2600	55 ton capacity		24	735	2,200	6,600	632	
	2700	80 ton capacity	↓	27	1,250	3,750	11,300	966	
	2750								
	2800	Self-propelled, 4 x 4, with telescoping boom, 5 ton	Ea.	6.25	185	560	1,675	162	
	2900	12-½ ton capacity	"	11.70	330	995	2,975	292.60	

17

Figure 4.2

The need to consider the maintenance overhead costs can be either casual or rigorous. While it is always important to find reductions in the costs of maintenance, the actual overhead costs are usually tracked only as a comparison with the original budget. Where work is done on a reimbursable basis, however, it is necessary to develop a maintenance overhead rate. This overhead rate is a pro-rated amount of the total cost of all indirect maintenance work.

Maintenance Management Salaries

Any employee within the maintenance organization who does not directly contribute to the actual performance of direct maintenance work must be included in the classification of *overhead costs*. Particular care must be taken to determine the extent that the first line supervisors, the shop foremen, are involved in indirect work. Depending on the size of the maintenance organization or individual shop, the shop foreman may spend a significant portion of his or her time in the performance of direct maintenance work. In other cases, the foreman's role is strictly supervisory. The shop foreman is involved in the "management" of the shop, ordering tools and supplies, tabulating work statistics, and developing work schedules for the employees. In either case, the exact amount of time dedicated to support rather than direct work should be tabulated, along with the attendant cost in wages. Since shop foremen are most frequently paid an hourly wage, this tabulation can be made in hours. Specific recommended procedures for tracking this split effort are included in Chapter 8, "Evaluating and Controlling the Maintenance Effort."

This split in duties should not be confused with the normal on-the-job supervision provided by a work leader or job foreman. These individuals are performing their supervisory duties in direct support of individual and specific maintenance projects. Costs for such supervision are *not* included as overhead costs. At managerial levels above the shop foreman, all personnel are considered overhead and their attendant salaries (and benefits) are overhead costs. The exact value of the salaries is known and is not subject to estimating.

Office and Shop Expenses

At the office level, all materials, supplies, forms, furniture, and utilities represent indirect maintenance costs. Office operating costs also include costs for reproduction equipment, telephones, mail, shipping and, if appropriate, utilities such as heat, water, and electricity. Figure 4.3 is a sample Office Overhead Budget Preparation Form, which includes the primary elements of office expenses.

Shop overhead costs include the costs of tools mentioned earlier in this chapter. Additionally, all costs incurred by the individual shop which do not directly relate to the completion of any single maintenance project are included in overhead costs. When developing the shop overhead budget, a major management decision must be made concerning the extent to which the maintenance manager will attempt to trace maintenance costs back to individual maintenance projects.

The simplest method for maintenance cost tracking is to have none at all. The total maintenance costs are simply totalled and no attempt is made to discern the cost of individual maintenance activities. On the opposite end of the spectrum would be a detailed accounting of time, materials, and overhead costs to the finest detail, enabling all costs to be related to individual projects. Such an elaborate system is cumbersome, time consuming, and provides little information from which significant

Maintenance Office Expenses
Budget Preparation Form

Expense Item	Annual Expense	Monthly Expense	Total Expense
Salaries			
Maintenance Manager	_____	_____	_____
Planner/Estimator	_____	_____	_____
Scheduler	_____	_____	_____
Work Receptionist	_____	_____	_____
Other			
_____	_____	_____	_____
_____	_____	_____	_____
_____	_____	_____	_____
Paperwork Support			
Copying Machine(s)	_____	_____	_____
Copying Machine Maintenance	_____	_____	_____
Copying Machine Paper	_____	_____	_____
Facsimile Machine	_____	_____	_____
Facsimile Machine Paper	_____	_____	_____
Envelopes, Mailing Cartons	_____	_____	_____
Stationery, Custom Letterhead	_____	_____	_____
Stationery, General Paper	_____	_____	_____
Office Supplies (pens, etc.)	_____	_____	_____
_____	_____	_____	_____
Mailing and Shipping			
Regular Postal Expense	_____	_____	_____
Overnight Mail	_____	_____	_____
Parcel Shipping	_____	_____	_____
Communication Expenses			
Phone System Maintenance	_____	_____	_____
Local Phone Bills	_____	_____	_____
Long Distance Phone Services	_____	_____	_____
Phone Line Charges	_____	_____	_____
Beeper System Charges	_____	_____	_____
Beeper Maintenance	_____	_____	_____
Computer Support Expenses			
Hardware & Hardware Support	_____	_____	_____
Software & Software Support	_____	_____	_____
Computer Paper, Disks, etc.	_____	_____	_____
Office Travel			
Local Travel Expenses	_____	_____	_____
Extended Travel & Per Diem	_____	_____	_____

Figure 4.3

reduction in maintenance costs can be made. A compromise between the two extremes is more reasonable. The overhead costs are collated by specific category, the total overhead cost determined, and then added as a fixed percentage overhead rate to any jobs for which total reimbursement is to be provided. Figure 4.4 is a sample Shop Overhead Summary Budget Form which includes typical indirect or overhead shop expenses.

Maintenance Control

The cost of maintenance control, or the system used to track and monitor the maintenance effort, is primarily the cost of the computer or manual system employed. However, there are many items within each system to be considered. The cost of a maintenance management computer system, for example, extends beyond the purchase or lease of the computer hardware. The license for the software must be paid both at initial installation and also periodically to ensure that proper support and updates are provided. The printing of standard forms to allow personnel to interact with the computer and the cost for computer paper used in printing the numerous reports must also be considered.

Unit Price Estimating Techniques

Before outlining the specific considerations for estimating the costs of the various types of maintenance activities, some basic methods of estimating must be explained. This section describes the particular techniques of *unit price estimating*. For any activity there is a requirement to expend some quantity of labor, to utilize some quantity of materials, and to employ some tools or equipment. If the exact quantities of each of these resources is known, the estimate is calculated simply. The sheer number of different maintenance activities precludes the development of sufficient historical data to describe the exact number of man-hours necessary for

Maintenance Office Expenses (continued) Budget Preparation Form			
Expense Item	Annual Expense	Monthly Expense	Total Expense
Training			
Seminars, Courses Tuition	_____	_____	_____
Seminar Travel & Per Diem	_____	_____	_____
Other Miscellaneous Expenses (itemize)			
_____	_____	_____	_____
_____	_____	_____	_____
_____	_____	_____	_____
_____	_____	_____	_____
_____	_____	_____	_____
_____	_____	_____	_____

Figure 4.3 (continued)

Maintenance Shop Overhead Expenses
Budget Preparation Form

Expense Item	Annual Expense	Monthly Expense	Total Expense
Craftsman Support			
Safety Equipment (glasses, etc.)	_____	_____	_____
Protective Clothing	_____	_____	_____
Hand Soap, Towels, etc.	_____	_____	_____
Training, Seminars Tuition	_____	_____	_____
Training Travel & Per Diem	_____	_____	_____
_____	_____	_____	_____
Tool Support			
Hand Tool Replacements	_____	_____	_____
Hand Tool Repair & Maintenance	_____	_____	_____
Power Tool Replacement	_____	_____	_____
Power Tool Repair & Maintenance	_____	_____	_____
Shop Equipment Replacement	_____	_____	_____
Shop Equipment Repair	_____	_____	_____
Tool Consumables (drill bits, sanding belts, saw blades, etc.)	_____	_____	_____
_____	_____	_____	_____
Shop Miscellaneous Materials			
Towels, Cleaning Rags	_____	_____	_____
Cleaning Fluids, Soaps	_____	_____	_____
Toilet Supplies	_____	_____	_____
_____	_____	_____	_____
_____	_____	_____	_____
Routine Consumable Replacement (See Note 1 below)			
_____	_____	_____	_____
_____	_____	_____	_____
Other Shop Expenses (See Note 2 below)			
_____	_____	_____	_____
_____	_____	_____	_____

Notes:
1. This category is used to purchase normal supplies which are consumed during the course of normal maintenance and repair work, but are used individually in such small quantities that accounting for use on each activity is not feasible. For example, lubricating grease, electrical tape, nails, screws, and O-rings, are purchased in bulk, but are used in very small amounts on any activity.
2. Some overhead expenses, such as beepers, and phone charges, are common to all shops and are carried as total Maintenance Department overhead costs. Department policy will determine which costs are collated and which are charged out at the shop level.

Figure 4.4

every task. Since many activities are similar in nature but vary in scope, an average level of productivity can be used. Using an average significantly decreases the number of activities for which individual productivity rates must be developed. For example, when a painter is painting large sections of a wall, he is able to paint at a specific rate for jobs of common, but not equal, size. Except for the smallest jobs where set-up and break-down times are longer than the time spent actually painting, the average rate can be applied to varying painting jobs to determine the man-hours required for each.

Predictable productivity rates are very helpful in estimating the costs for widely varying maintenance tasks. The estimator can develop, from knowledge and observation, the average productivity rates for various groups of tasks. Where data is lacking or insufficient to predict an average rate of production, the estimator should consult a published cost guide such as *Means Facilities Cost Data*. This and similar cost guides compile the productivity and cost data for thousands of individual tasks in facility construction, remodeling, and maintenance. The data is updated annually to reflect changes in the costs for labor, material, or equipment. In order to be universally useful, the data represents an average wage, material cost, equipment cost, or productivity rate from many geographic areas. Correction factors are offered to convert the standard costs to more appropriate local wages and material prices. Where the wage rate or material cost used by the cost guide is significantly different from the actual wages or prices paid by the using organization, the cost guide remains useful in the area of productivity. While wages vary greatly geographically, the individual worker productivities are more constant. (*Means Man-Hour Standards* is another helpful source of productivity information.)

The estimator chooses the appropriate sources for costs in an estimate for each individual activity. Where historical data is adequate and applicable, it is used and updated. When the exact hourly costs for labor are known, the estimator uses those figures. If hourly wages and productivity rates are unknown, a cost guide can be used. If a cost guide is to be used, the estimator must understand the methods for developing the data within the cost guide. Figures 4.5 and 4.6 show the organization and use of *Means Facilities Cost Data, 1988*, taken from the annual cost guide. The estimator should consult the cost guide to determine the applicability of specific wage rates to the maintenance organization. For unit price estimates, the estimator uses a specific hourly productivity rate for a task.

Direct Maintenance Activity Costs

Cost estimates for direct maintenance activities are prepared for several reasons, primarily to provide a yardstick against which the actual progress of the work can be measured. The key element in preparing an accurate estimate is not the amount of the labor hourly rate, but rather the assumed productivity rate. Therefore the impact of productivity of individual maintenance activities must be understood. The predicted productivity rate must be reasonable.

When labor costs make up the predominant portion of the cost of a given activity, emphasis is directed to actual supervision of the employees. Conversely, if the costs are primarily generated by the purchase of materials and parts, increased emphasis is placed on obtaining competitive material prices. The following section describes the prevailing proportions of labor, material, and tools for each activity type. These proportions indicate both the need for emphasis on accurate estimating and the areas in which increased management supervision may result in the greatest reductions in maintenance costs.

HOW TO USE THIS BOOK

HOW THE BOOK IS ARRANGED

This book is divided into four sections: Unit Price, Assemblies, Reference and an Appendix.

Unit Price Section: All cost data has been divided into the 16 divisions of the Construction Specifications Institute's (C.S.I.) MASTERFORMAT plus a S.F. (square foot) and C.F. (cubic foot) Cost division (17). A listing of these divisions and an outline of their subdivisions is shown in the Table of Contents page at the beginning of the Unit Price Section.

Numbering Each unit price line item has been assigned a unique 10 digit code. A graphic explanation of the numbering system is shown on the "How to Use Unit Price" page.

Descriptions Each line item number is followed by a description of the item. Sub-items and additional sizes are indented beneath appropriate line items. The first line or two after the main (bold face) item often contain descriptive information that pertains to all line items beneath this bold face listing.

Crew The "Crew" column designates the typical trade or crew to install the item. When an installation is done by one trade and requires no power equipment, that trade is listed. For example, "2 Carp" indicates that the installation is done with 2 carpenters. Where a composite crew is appropriate, a crew code designation is listed. For example, a "C-2" crew is made up of 1 foreman, 4 carpenters, 1 laborer plus power tools. All crews are listed at the beginning of the Unit Price Section. Costs are shown both with bare labor rates, and with the contractor's overhead and profit. For each, the total crew cost per eight-hour day and the composite cost per man-hour are listed.

Crew Equipment Cost The power equipment required for each crew is included in the crew cost. The daily cost for crew equipment is based on dividing the weekly bare rental rate by 5 (number of working days per week), and then adding the hourly operating cost times 8 (hours per day). This "Crew Equipment Cost" is listed in Division 016.

Daily Output To the right of every "Crew" code listing, a "Daily Output" figure is given. This is the number of units that the listed crew will install in a normal 8-hour day.

Man-hours The column following "Daily Output" is "Man-hours". This figure represents the man-hours required to install one "unit" of work. Unit man-hours are calculated by dividing the total daily crew hours (as seen in the Crew Tables) by the Daily Output.

Unit To the right of the "Man-hour" column is the "Unit" column. The abbreviated designations indicate the unit upon which the price, production and crew are based. See the Appendix for a complete list of abbreviations.

Material The first column under the "Bare Cost" heading lists the unit material cost for the line item. This figure is the "bare" material cost with no overhead and profit allowances included. Costs shown reflect national average material prices for January of the current year and include delivery to the jobsite.

Labor The second "Bare Cost" column is the unit labor cost. This cost is derived by dividing the daily labor cost by the daily output. The wage rates used are listed on the inside back cover.

Equipment The third "Bare Cost" column lists the unit equipment cost. This figure is the daily crew equipment cost divided by the daily output.

Total The last "Bare Cost" column lists the total bare cost of the item. This is the arithmetic total of the three previous columns: "Material", "Labor", and "Equipment".

Total Incl. O&P The figure in this column is the sum of three components: the bare material cost plus 10%; the bare labor cost plus overhead and profit (per the labor rate table on the inside back cover); and the bare equipment cost plus 10%. A sample calculation is shown on the "How to Use Unit Price" page.

Division 17 contains Square Foot and Cubic Foot costs for 59 different building types. These figures include contractor's overhead and profit but do not include architectural fees or land costs.

Assemblies Section: This section uses an "Assemblies" format grouping all the functional elements of a building into 12 "Uniformat" Construction Divisions.

At the top of each "Assembly" cost table is an illustration, a brief description, and the design criteria used to develop the cost. Each of the components and its contributing cost to the system is shown.

For a complete breakdown and explanation of a typical "Assemblies" page, see "How to Use Assemblies Cost Tables" at the beginning of this section.

Material These cost figures include a standard 10% markup for "handling". They are national average material costs as of January of the current year and include delivery to the jobsite.

Installation The installation costs include labor and equipment, plus a markup for the installing contractor's overhead and profit.

Reference Section: Following the items in the "Unit Price" pages, there are frequently found large numbers in circles. These numbers refer the reader to data in this Reference Section. This material includes estimating procedures, alternate pricing methods, technical data and cost derivations. This section also includes information on design and economy in construction.

Appendix: Included in this section are Historical and City Cost Indexes, a list of abbreviations and a comprehensive index.

Historical Cost Index This index provides annual data to adjust construction costs over time.

City Cost Indexes These indexes provide data to adjust the "national average" costs in this book to 162 major cities throughout the U.S. and Canada.

Abbreviations/Index A listing of the abbreviations used throughout this book, along with the terms they represent is included. Following the abbreviations list is an index for all sections.

PROJECT SIZE

The book is aimed primarily at industrial, and commercial projects costing $5,000 to $500,000. With reasonable exercise of judgment the figures can be used for any building project but do not apply to civil engineering structures such as bridges, dams, highways or the like.

ROUNDING OF COSTS

In general, all unit prices in excess of $5.00 have been rounded to make them easier to use and still maintain adequate precision of the results. The rounding rules are as follows:

Price from $5.01 to $20.00 rounded to the nearest 5¢

Price from $20.01 to $100.00 rounded to the nearest $1

Price from $100.01 to $1,000.00 rounded to the nearest $5

Price from $1,000.01 to $10,000.00 rounded to the nearest $25

Price from $10,000.01 to $50,000.00 rounded to the nearest $100

Price over $50,000.01 rounded to the nearest $500

Figure 4.5

HOW TO USE UNIT PRICE PAGES

Important
Prices in this section are listed in two ways: as bare costs and as costs including overhead and profit of the installing contractor. In most cases, if the work is to be subcontracted, it is best for a general contractor to add an additional 10% to the figures found in the column titled **"TOTAL INCL. O&P"**.

Unit
The unit of measure listed here reflects the material being used in the line item. For example: bracing is defined in linear feet (L.F.).

Productivity
The daily output represents typical total daily amount of work that the designated crew will produce. Man-hours are a unit of measure for the labor involved in performing a task. To derive the total man-hours for a task, multiply the quantity of the item involved times the man-hour figure shown.

Line Number Determination
Each line item is identified by a unique ten-digit number.

MASTERFORMAT
Division
061 104 0200
Subdivision

MASTERFORMAT
Mediumscope
061 100
061 **104** 0200
Major Classification

061 104 **0200**
Individual Line Number

Description
The meaning of this line shows wood let-in bracing with 1" x 6" boards will be installed by an F-1 crew at a rate of .035 man-hours per linear foot. Stud spacing is 24" on center.

(86) **Circle Reference Number**
These reference numbers refer to charts, tables, estimating data, cost derivations and other information which may be useful to the user of this book. This information is located in the Reference Section of this book.

Crew F-1

Crew No.	Bare Costs		Incl. Subs O & P		Cost Per Man-hour	
	Hr.	Daily	Hr.	Daily	Bare Costs	Incl. O&P
Crew F-1						
1 Carpenter	$21.20	$169.60	$33.60	$268.80	$21.20	$33.60
Power Tools		6.80		7.50	.85	.93
8 M.H., Daily Totals		$176.40		$276.30	$22.05	$34.53

Bare Costs are developed as follows for line no. **061-104-0200**
Mat. is **Bare Material Cost ($.19)**
Labor for Crew F1 = Man-hour Cost **($21.20)** × Man-hour Units **(.035) = $.74**
Equip. for Crew F1 = Equip. Hour Cost **($.85)** × Man-hour Units **(.035) = $.03**
Total = **Mat. Cost ($.19)** + **Labor Cost ($.74)** + **Equip. Cost ($.03) = $.96**
(**Note:** When a Crew is indicated Equipment and Labor costs are derived from the Crew Tables. See example above.)

Total Costs Including O&P are developed as follows:
Mat. is **Bare Material Cost** + 10% = **$.19** + **$.02** = **$.21**
Labor for Crew F1 = Man-hour Cost **($33.60)** × Man-hour Units **(.035) = $1.17**
Equip. for Crew F1 = Equip. Hour Cost **($.93)** × Man-hour Units **(.035) = $.03**
Total = **Mat. Cost ($.21)** + **Labor Cost ($1.17)** + **Equip. Cost ($.03) = $1.41**
(**Note:** Where a crew is indicated, Equipment and Labor costs are derived from the Crew Tables. See example at top of this page. **"Total"** line costs are rounded.)

061 | Rough Carpentry

061 100 | Wood Framing

			CREW	DAILY OUTPUT	MAN-HOURS	UNIT	MAT.	LABOR	EQUIP.	TOTAL	TOTAL INCL O&P	
102	0011	BLOCKING										102
	2600	Miscellaneous, to wood construction										
	2620	2" x 4"	F-1	.17	47.060	M.B.F.	375	1,000	40	1,415	2,050	
	2660	2" x 8"	*	.27	29.630	*	400	630	25	1,055	1,475	
	2720	To steel construction										
	2740	2" x 4"	F-1	.14	57.140	M.B.F.	375	1,200	49	1,624	2,375	
	2780	2" x 8"	*	.21	38.100	*	400	810	32	1,242	1,750	
104	0010	BRACING Let-in, with 1" x 6" boards, studs @ 16" O.C.	F-1	150	.053	L.F.	.19	1.13	.05	1.37	2.05	104
	0200	Studs @ 24" O.C.		230	.035	*	.19	.74	.03	.96	1.41	
106	0010	BRIDGING Wood, for joists 16" O.C., 1" x 3"		130	.062	Pr.	.28	1.30	.05	1.63	2.44	106
	0100	2" x 3" bridging		130	.062		.45	1.30	.05	1.61	2.64	
	0300	Steel, galvanized, 18 ga., for 2" x 10" joists at 12" O.C.	1 Carp	130	.062		.69	1.30		1.99	2.83	
	0400	24" O.C.		140	.057		1.10	1.21		2.31	3.13	
	0900	Compression type, 16" O.C., 2" x 8" joists		200	.040		1.05	.85		1.90	2.50	
	1000	2" x 12" joists		200	.040		1.15	.85		2	2.61	
108	0010	FRAMING, LIGHT Average for all light framing	F-2	1.05	15.240	M.B.F.	380	325	12.95	717.95	945	108
	2300	Joists, fir,2" x 4"		.85	18.820		375	400	16	791	1,075	
	3300	Mud sills, redwood, construction grade, 2" x 4"		.60	26.670		875	565	23	1,463	1,875	
	3400	Nailers, treated, 2" x 4" to 2" x 8" wood construction		.75	21.330		515	450	18.15	983.15	1,300	
	4300	Post, columns & girts, 4" x 4"		.52	30.770		560	650	26	1,236	1,675	
	5400	Roof cants, split 4" x 4"		6.50	2.460	C.L.F.	91	52	2.09	145.09	185	
	5500	Split 6" x 6"		6	2.670	*	207	57	2.27	266.27	320	
	6100	Rough bucks, treated, for doors or windows, 2" x 6"		400	.040	L.F.	.52	.85	.03	1.40	1.95	
	6500	Sills, 4" x 4"		.78	20.510	M.B.F.	645	435	17.45	1,097	1,425	
	6800	Sleepers on concrete, treated, 1" x 2"		2,350	.007	L.F.	.09	.14	.01	.24	.34	
	7300	Stair stringers, fir, 2" x 10"		.30	53.330	M.B.F.	450	1,125	45	1,620	2,325	
110	0010	FRAMING, BEAMS & GIRDERS										110
	1000	Single, 2" x 6"	F-2	700	.023	L.F.	.38	.48	.02	.88	1.21	
	1020	2" x 8"		650	.025		.54	.52	.02	1.08	1.44	
	1040	2" x 10"	(86)	600	.027		.75	.57	.02	1.34	1.75	
	1060	2" x 12"		550	.029		.86	.62	.02	1.50	1.95	
	1080	2" x 14"		500	.032		1.08	.68	.03	1.79	2.30	

Figure 4.6

Housekeeping

Housekeeping activities are predominantly labor-intensive. Janitorial workers, for example, consume few materials or supplies, employ some labor-saving devices, and work to a specified quality level in a specified area of the facility. Outside work in landscaping and gardening are more task-oriented but, except large scale lawn mowing, are also predominantly labor intensive. Estimates in the housekeeping area are generally prepared only for specific tasks which have definitive predicted times of occurrence, are not recurrent on a daily basis, and are directly measurable for completion.

Productivity: The method for determining the staff required for housekeeping activities is often the result of both estimating and observing. Estimates for productivity rates for various tasks can be derived from published cost data such as *Means Facilities Cost Data, 1988*, as shown in Figure 4.7.

However, since housekeeping tasks are repeated with great regularity, a reasonable productivity rate can be established from historical observations. Labor-saving devices offer the greatest potential for improving productivity. (Evaluating potential labor-saving devices is discussed in Chapter 5, "Cost-Based Maintenance Decisions.") Some degree of improved quality and productivity can be achieved through management direct supervision and frequent feedback to the employee, with no increase in maintenance costs. Once the full limits of an employee's capabilities are reached, only step-wise improvements in productivity are possible. That is, most housekeeping tasks are performed in the course of a full workday with specific duties carried out within certain work areas. A janitor is responsible for cleaning one or more floors in an office building. When that workload proves to be too much, it is difficult to augment that individual's efforts with part-time help. Another full-time employee is added yielding a net decrease in work per salary dollar. In view of this difficulty in matching work force to workload, the normal tendency is to push for higher productivity levels until quality drops below a minimally acceptable level. At that point the added staff provides an increased quality of service, which will be appreciated by the customers.

An example of the method for estimating the cost of a housekeeping activity is shown in Figure 4.8. This particular activity involves vacuuming 12,000 square feet of general office space.

In selecting the appropriate line item from Figure 4.7, the choice was made to use line number 018-090-0550, "Carpet cleaning, vacuum, dry pick-up, obstructed." If the space to be vacuumed had been predominantly corridors or lobbies, line 018-090-540, the line item above, "Unobstructed," would have been a better choice. Several pieces of information are included within each specific line item. The "Crew" is shown as "1 Clab," a common laborer. The "Daily Output" is 22 units per day. The man-hours per unit is .364 and the units used on this line are M.S.F., thousand square feet. No costs are shown for material or equipment. Although it is obvious that a vacuum is needed, the unit cost of a vacuum is insignificant in comparison to the labor costs. The *Bare Cost* is a simple addition of the bare labor, material, and equipment costs. The *Total Including Overhead and Profit* is the total of the Bare Costs, Overhead, and Profit (added as a percentage). This higher cost is useful when the work being estimated is performed under contract by an outside firm or person.

			DAILY	MAN-		BARE COSTS				TOTAL	
018 000	**Facilities Maintenance**	CREW	OUTPUT	HOURS	UNIT	MAT.	LABOR	EQUIP.	TOTAL	INCL O&P	
080 2080	Clear glass partition, 8 sq. ft. per unit	1 Clab	3	2.670	M.S.F.		44		44	70	080
2090	Opaque, 20 sq. ft. per unit	"	3	2.670	"		44		44	70	
2100	Alternate pricing method by window area to be washed										
2110	Minimum productivity	1 Clab	1	8	M.S.F.		130		130	210	
2120	Average productivity		2.50	3.200			53		53	84	
2130	Maximum productivity	↓	4	2	↓		33		33	52	
2200	Weather stripping see Division 087-306										
2210											
090 0010	FINISHES, FACILITIES MAINTENANCE										090
0100	Ceiling maintenance										
0120	Acoustical tile cleaning, chemical				S.F.					.25	
0130	Clean and apply accoustical coating ⑲⑥				"					.60	
0400	Floor maintenance										
0500	Carpet fiber sealant, anti-soilant, commercial				S.F.					.25	
0510	Residential				"					.35	
0530	Carpet cleaning, vacuum, dry pick-up										
0540	Unobstructed	1 Clab	30	.267	M.S.F.		4.41		4.41	7	
0550	Obstructed		22	.364			6		6	9.55	
0560	Wet pick-up, unobstructed		21	.381			6.30		6.30	10	
0570	Obstructed	↓	14	.571	↓		9.45		9.45	15	
0580	Steam clean, minimum		9,760	.001	S.F.	.03	.01		.04	.05	
0590	Maximum cost	↓	3,250	.002	"	.04	.04		.08	.11	
0700	Composition, resilient or wood flooring										
0720	Dust mop, unobstructed	1 Clab	60	.133	M.S.F.		2.21		2.21	3.50	
0730	Obstructed		35	.229			3.78		3.78	6	
0740	Damp mop, unobstructed		26	.308		.40	5.10		5.50	8.50	
0750	Obstructed		16	.500		.40	8.30		8.70	13.55	
0780	Hand scrub, unobstructed		1.80	4.440		.40	74		74.40	115	
0790	Obstructed		1.30	6.150		.40	100		100.40	160	
0800	Machine scrub, unobstructed		17	.471		.40	7.80		8.20	12.80	
0810	Obstructed		12	.667		.40	11.05		11.45	17.95	
0820	Machine polish, unobstructed		28	.286		7.30	4.73		12.03	15.50	
0830	Obstructed	↓	14	.571	↓	7.30	9.45		16.75	23	
0850	Refinish old wood floors, minimum	1 Carp	400	.020	S.F.	.43	.42		.85	1.15	
0860	Maximum cost	"	130	.062	"	.65	1.30		1.95	2.78	
0870	Strip & rewax/polish, unobstructed ⑲⑥	1 Clab	4	2	M.S.F.	7.65	33		40.65	61	
0880	Obstructed		3	2.670		7.65	44		51.65	78	
0890	Sweeping, unobstructed		47	.170			2.82		2.82	4.46	
0900	Obstructed	↓	35	.229	↓		3.78		3.78	6	
0950	Painting concrete or wood floors, see division 099-220										
1200	Paint removal, see division 099-902										
1300	Scrape after fire damage, see division 099-906										
2000	Wall maintenance										
2400	Washing enamel finish walls with mild cleanser	1 Clab	16	.500	M.S.F.	.25	8.30		8.55	13.40	
100 0010	SPECIALTIES, FACILITIES MAINTENANCE										100
0100	Bathroom accessories										
0120	Accessory, replacement see Division 108-204										
0130	All purpose cleaner, concentrate				Gal.	6.60			6.60	7.25	
0150	Clean mirror, 36" x 24"	1 Clab	840	.010	Ea.	.01	.16		.17	.26	
0160	48" x 24"		630	.013		.01	.21		.22	.34	
0170	72" x 24"	↓	420	.019	↓	.01	.32		.33	.51	
0200	General cleaning of fixtures (basins, water closets, urinals)										
0210	Including shelves, partitions and dispenser servicing	1 Clab	80	.100	Fixture		1.66		1.66	2.62	
1500	Partition replacement, hospital, office, toilet see Division 106-300										
110 0010	ARCHITECTURAL EQUIPMENT, FACILITIES MAINTENANCE										110
0100	Bird control needle strips										
0110	Anchor clip mounted	2 Clab	500	.032	L.F.	4.25	.53		4.78	5.50	
0120	Adhesive mounted	"	550	.029	"	4.60	.48		5.08	5.80	

24

Figure 4.7

The final result of the sample estimate is a predicted cost of $72.00 and a predicted activity duration of four and one third man-hours. At this point, the estimator may choose to temper the estimate with local judgment. In this case the janitor may need to go to another section of the facility to pick up the vacuum and, accordingly, additional time (and its attendant cost) must be added to the estimate. The estimate, once prepared, may be used for several purposes, all of which depend on the accuracy of the estimate. The maintenance supervisor may use the data to determine the desired frequency for vacuuming the offices. The housekeeeping foreman uses the estimated duration in man-hours to determine the appropriate crew size and then to schedule the daily tasks for his personnel. The shop foreman may, again, use the estimate to measure the progress and performance of the individual assigned the vacuuming duties.

Materials, Parts, and Supplies: The predominant housekeeping materials and supplies consumed are cleaning materials and consumable paper products such as toilet paper or paper towels. Housekeeping materials and supplies are normally used in widely varying quantities. For this reason, it is of little use to define the exact amount of material to be expended on each task. A stock of supplies is initially purchased and then the average consumption rate is determined. Replacement supplies are ordered as needed. This total cost for supplies is carried as a separate budget item under the general housekeeping shop category, rather than tracking the cost for each use.

Maintenance Activity: Vacuum Office Spaces
(12,000 Square Feet)

Calculations:
  Activity Cost:
    From Figure 4.5, Line 018-090-0550
      Unit Cost is $6.00 per 1000 sq. ft.

$$\text{Total Cost} = 12,000 \times \frac{\$6.00}{1000}$$
$$= \$72.00$$

  Activity Duration:
    Daily Output = 22,000 sq. ft. per 8 Hour Day
    Activity = 12,000 sq. ft.

$$\text{Duration} = \frac{12,000}{22,000} = .545 \text{ Days}$$
$$= 4 \text{ Hours 22 Minutes}$$

Figure 4.8

Tools and Equipment: The cost of providing any tool includes both the purchase (or replacement) price of the tool and the cost associated with its daily operation. The operating costs include any necessary fuel, lubricants, filters, or parts, such as sandpaper for sanders or drill bits for electric drills. The labor required to operate the tool is not included with tool costs. The ownership and operating costs for housekeeping tools are low in comparison to the labor cost associated with using the tool. For this reason, the cost of housekeeping tools is rarely included in individual activity cost estimates. The costs of tools and equipment are tracked through a *small tools allowance* or as *planned tool replacement line items* within the housekeeping budget. Exceptions are those few housekeeping activities which require the operation of expensive and complex equipment. In these cases, the cost of the machine or equipment is justified by the anticipated savings in labor. Hence, the labor cost for the use of this machine is not the predominant cost, and the equipment operating cost should be included in the overall activity cost.

General Maintenance

As noted in Chapter 1, general maintenance falls into two categories: *infrequent housekeeping* and *nuisance work*. Infrequent housekeeping activities are those which must be performed from time to time, not continuously. Weeding planting beds, waxing floors, sweeping streets and parking lots, and steam cleaning carpets are typical examples. Although these tasks are performed somewhat regularly, they are distinct events which require formal scheduling and individual estimating. Nuisance work, also not regularly scheduled, includes minor adjustment or repair to the parts, fixtures, and hardware of the entire facility. They are discrepancies which are generally noted and reported by the facility user. Unclogging a toilet, adjusting a door closer, tightening loose screws, and replacing a broken ceiling tile are typical nuisance work items.

Labor Cost: The two types of general maintenance activities differ in the required proportions of labor, material, and equipment. The distribution for infrequent housekeeping activities is similar to that for recurring housekeeping tasks. While still labor intensive, infrequent housekeeping may also include some equipment costs. For instance, steam cleaning carpets requires the ownership or rental of steam-cleaning equipment. Unlike the vacuum in the housekeeping example, steam-cleaning equipment is much more costly to purchase and to operate, since it requires the use of special soaps and defoamers. The method for estimating the costs of these activities is similar to that for repetitive housekeeping tasks, with the added equipment cost. The predicted number of man-hours required to complete the project is often also the project duration. Increasing the crew can only shorten the activity if sufficient specialized equipment is available.

The nuisance work tasks are usually completed within a short period of time. This introduces a new variable into the predicted duration of an activity. The length of time required to gather the proper tools and parts, travel to the work site, travel back to the shop, and put away the tools is often several times longer than the duration of the actual repair or adjustment. Figure 4.9 shows the cost and duration for the simple general maintenance activity of replacing a single fluorescent light tube.

In this example extract from a page of *Means Facilities Cost Data, 1988,* an electrician can install several light bulbs in an hour at a cost of $7.25 each. The addition of the assumed transit times increases the unit cost by $7.07 when only one bulb is changed. (It should be noted that the cost guide assumes that the bulb will be changed by an electrician. While this may be the case where union work rules require that only an electrician can perform this task, a non-union maintenance organization may use a maintenance mechanic or janitor to change the bulbs at much lower hourly wages.) When two bulbs are changed in nearby locations the additional cost is only the $7.25 shown in the cost guide and the average per bulb cost drops to $10.79. When three are changed, the cost drops to $9.61 each. This example highlights the desirability of delaying the completion of nuisance items until several tasks can be performed in a single area by a single worker.

Maintenance Activity: Replace One Fluorescent Light Tube

Calculations:

Pick up Ladder, Tube, Walk to Location	= .15 Hours
Replace Tube (Line Item 018-160-2515)	= .133 Hours
Return Ladder, Dispose Old Tube	= .15 Hours
	Total = .433 Hours

Labor Cost = .433 Hours @ $23.60	= $10.22
Material (Line Item 018-160-2515)	= $ 4.10
Equipment (Negligible)	= $ 0.00
	Total Activity Cost = $14.32

018 | Maintenance

		018 000	Facilities Maintenance	CREW	DAILY OUTPUT	MAN-HOURS	UNIT	BARE COSTS				TOTAL INCL O&P	
								MAT.	LABOR	EQUIP.	TOTAL		
160	0740		Replace lead connection	1 Elec	4	2	Ea.		47		47	71	160
	0800		Control device, install		5.70	1.400			33		33	50	
	0810		Disassemble, clean and reinstall		7	1.140			27		27	41	
	0820		Replace		10.70	.748			17.65		17.65	27	
	0830		Trouble shoot	▼	10	.800	▼		18.90		18.90	28	
	0900		Demolition, for electrical demolition see Division 020-708										
	1000		Distribution systems and equipment install or repair a breaker										
	1010		In power panels up to 200 amps	1 Elec	7	1.140	Ea.		27		27	41	
	1020		Over 200 amps		2	4			94		94	140	
	1030		Reset breaker or replace fuse		20	.400			9.45		9.45	14.20	
	1100		Megger test MCC (each stack)		4	2			47		47	71	
	1110		MCC vacuum and clean (each stack)		5.30	1.510			36		36	54	
	2500		Lighting equipment, replace road light		3	2.670		670	63		733	830	
	2510		Fluorescent fixture		7	1.140		54	27		81	100	
	2515		Relamp (flour.) facility area ea. tube		60	.133		4.10	3.15		7.25	9.25	
	2518		Fluorescent fixture, clean (area)		44	.182			4.29		4.29	6.45	
	2520		Incandescent fixture		11	.727		41	17.15		58.15	71	
	2530		Lamp (incadescent or flourescent)		60	.133		4.10	3.15		7.25	9.25	

Figure 4.9

Cost of Materials: The cost of materials and supplies is easily determined for general maintenance activities. The spare parts utilized in these activities, such as light bulbs or faucet washers, are usually available from several different manufacturers, and pricing for these items is competitive. If a facility maintenance manager finds that the spare parts used in general maintenance activities are excessively expensive because of a restriction to a single manufacturer, he may consider replacing the unit with one which can be repaired more cheaply or which is simply more reliable.

Tools and Equipment: The cost of specialized equipment used for infrequent maintenance items is often of such a magnitude that it forces the maintenance manager to question whether or not the particular task should be performed in-house. For example, an activity such as steam cleaning could be performed by an outside contractor. The contractor who specializes in steam cleaning carpets and upholstery utilizes the expensive equipment on a continual basis and may be able to operate such equipment at a lower cost per hour than could a maintenance organization using this equipment infrequently. Another alternative is rental of such equipment.

The tool and equipment costs for minor repairs and adjustments are not significant since common hand tools are all that are normally required. Such hand tools are not charged to individual jobs but are carried as a single line item in the shop budget.

Preventive Maintenance

Preventive maintenance (PM) activities are the most predictable tasks of all maintenance work. The specific requirements for each task are clearly prescribed. The quantity of material used is consistent. The tools utilized remain constant. The task is repeated at regular intervals. These characteristics combine to make cost estimating both easy and reliable for preventive maintenance activities.

Productivity: A properly trained and equipped mechanic can perform a preventive maintenance task on a piece of equipment in a definite period of time, no more, no less. There are little options available to a manager to improve the production rate of a worker performing a preventive maintenance task. In order to reduce the overall cost of performing preventive maintenance, the manager should concentrate on efficient scheduling to minimize the transit time between tasks. This may require some delaying or premature performance of preventive maintenance on certain equipment to bring the scheduled accomplishment in sequence with nearby and similar equipment. When identical preventive maintenance is performed on several pieces of similar equipment, the craftsman gains some small but predictable proficiency at each step in the preventive maintenance activity. This increase in productivity will either carry over to the next repetition of the task, or, when the time between successive preventive maintenance tasks is large, the gained proficiency will be quickly restored as the worker starts the next series of similar tasks.

Some variance in time for performing identical preventive maintenance activities occasionally occurs. Minor problems with disassembly of a part or the discovery of a part which requires replacement may extend completion time beyond the estimated duration. Since preventive maintenance tasks often require that the equipment be de-energized, the worker might have to wait until the appropriate facility user can accommodate a loss of the use of the equipment. Transformer or high voltage switching gear maintenance requires that all circuits be de-activated. Since such disruptions are not easily accommodated by the facility user, this type of preventive maintenance task should be scheduled outside of normal work hours.

The estimated cost and duration for each preventive maintenance task is determined in the same manner as shown in previous examples. If there are no historical records (such as when a new piece of equipment is serviced for the first time) the estimate may be prepared using data from cost guides. After each particular activity is performed several times, however, the historical data is far more accurate than that which might be calculated otherwise. When the quantity of identical preventive maintenance activities is great, there may be some potential savings in job duration obtained through time-motion studies of the preventive maintenance mechanic. Through direct observation, coaching, and instruction, the mechanic may be able to achieve small but noticeable savings in time on individual activities which will become significant when applied time after time.

Materials, Parts, and Supplies: The materials required for each preventive maintenance task are spelled out explicitly by the manufacturer's recommended servicing procedures. If air handling units are to be cleaned, lubricated, the belts checked, and filters changed semi-anually, the exact number of filters required can be procured in bulk well in advance. The cost of that filter can be carried as a part of the individual preventive maintenance cost. Lubricants, however, are generally purchased in bulk and carried as a single line item in the servicing shop's budget since the consumption of lubricants for each unit serviced is low.

Tools and Equipment: Special tools and equipment necessary for preventive maintenance are unavoidable costs. The maintenance manager may locate a particular tool, such as a vibration sensor, which will enhance the mechanic's ability to diagnose potential problems. The manufacturer of such a tool may claim that the tool pays for itself in a short time by predicting potentially costly equipment failures. The maintenance manager should be wary of such claims and should carefully evaluate the potential savings of such diagnostic aids. For instance, if the failure which the tool is supposed to predict has never occurred in the facility's history, the tool should only be considered if it may reduce labor costs.

Repair

Repair activities are somewhat more complex than general maintenance tasks, usually involving a significant repair or overhaul of the major component of a system. Repair work begins when a problem, such as a lack of air conditioning in part of the building, is reported. The appropriate servicing mechanic is dispatched to the equipment room or problem area to begin an investigation into the problem. If a major element of the system is found to be in need of repair, the maintenance supervisor should be advised immediately. The supervisor must then decide whether to repair or replace the component. In an ideal situation, the problem would be analyzed in the manner described in Chapter 5, "Cost Based Maintenance Decisions," and the decision would be dictated by that analysis. In reality, however, the need to restore the system to operation, coupled with foreseen delays in purchasing and receiving a replacement part, may dictate that the component be repaired. Such repairs usually involve the removal of a component (such as a motor, compressor, valve, engine, control panel, or switch) which is taken back to the maintenance shop where appropriate tools and more suitable working areas are available. There the component is overhauled and restored to working condition. It is then returned to its original location, re-installed, and system operation is restored.

Repair projects are usually urgent projects. The work is essential to restoring operation of a system which is needed to carry on the routine operations of the facility. For this reason, overtime for repair projects is used as liberally as necessary. Low productivity in repair projects is normally not a problem. The workers, knowing the importance of restoring the system to operation, pursue the repairs with energy and diligence. Only casual supervision is usually necessary. A formal estimate is prepared only for the purpose of comparing repair with replacement. Thereafter, the workers proceed until repairs are completed, parts are used as needed, and any necessary equipment is employed. These are done without major concern for monitoring costs. Thus, the total cost of the repair emerges after the repairs are completed. Costs should not be completely disregarded, however, as productivity is a measure of output per dollar spent. If the workers are performing at optimal work rates, the area to monitor for excessive costs is in the staffing of the repair. It is often a policy to divert several workers from normal duties to the repair. As the scope of the repair emerges, the crew should be reduced to the absolute minimum. If there is any action which reduces productivity in repair projects, it is the tendency to assign more personnel to the task than necessary.

If a repair project is not an emergency, the attention to cost estimating is more important. When time allows the preparation of an accurate estimate, the repair versus replacement decision can be more realistic. The estimate is also used as the measure of progress as the repairs proceed. Figure 4.10 shows a method for preparing an estimate for a repair project.

The estimate shown in the example is similar to the estimates for general maintenance; the primary difference is the requirement that the failed component be moved twice—once from its installed location to the shop, then again from the shop to the site for reinstallation. Besides the implied costs of transporting the component, the cost of maintaining the shop space to complete the repair makes up a significant portion of overhead costs. If this overhead cost becomes unwieldy, a decision may be made that the major repair of certain components is done under contract with an outside service organization at the contractor's site. For example, motors are routinely sent out for rewinding, compressors sent to a local company for rebuilding, control panels are sent back to the manufacturer rather than being repaired in-house.

Replacement

The alternative to repair is replacement. As noted in the previous section, the decision to replace versus repair is mainly a cost decision. Replacement projects also present an opportunity to make reductions in the long-term operating costs for the facility. When a component has reached the end of its useful life and rebuilding or repair is not feasible, the replacement should provide equal or better service. Immediate replacement with an identical component should, however, not be automatically assumed.

Productivity: The replacement of a system component or entire system is similar to the initial installation of the system during facility construction. Difficulties may arise when it comes to fitting the replacement component into an existing system. The design engineer should have anticipated the replacement. The removal of a motorized valve in a piping system is eased by the presence of a bypass loop, isolation valves, and flanged or "union" fittings on either side of the valve. The replacement of some equipment, however, requires dismantling parts of the facility. Removal of doors, erection of special rigging to lift heavy items, and even the removal of walls to allow movement of large items may be necessary. Each of these considerations represents an impedence to smooth productivity. This may mean that the cost of the replacement is higher than that of new construction.

When estimating a replacement project, cost associated with the duration necessary to fit a new part to the existing system and location should be accurately estimated. The shop foreman or individual craftsman responsible for the repair can usually provide an estimate of the time required to compensate for these difficulties. The cost estimating for the initial installation can be prepared using cost guides such as *Means Facilities Cost Data*. Figures 4.11 and 4.12 are examples of cost estimates prepared for the initial installation and for replacement, respectively, of an air handling unit.

Repair Project Cost Estimate

Maintenance Activity: Overhaul and Rewind 7HP Electric Motor Located in Restricted Area

Work Breakdown and Cost Estimate:

Item	Reference*	Material	Labor	Man-Hours
Remove Motor	163-510-5170		$125	5.33
Disassemble, Clean	018-160-3040		94	4
Rewind Motor	018-160-3110		125	5.33
Reinstall	163-510-5170		125	5.33
Reconnect	160-275-0100	$3.50	34	1.4
	Subtotal	$3.50	$503	
	Total	$506.50		

Activity Duration:

Item	Total Man Hours	Crew Size	Item Duration
Remove	5.33	2	2.67
Clean, Rewind	9.33	1	9.33
Reinstall	6.78	2	3.39
		Total	15.39

Duration Start to Complete = 15 Hours 24 Minutes = 16 Hours

Note: Task is 99% labor.

*Line item from *Means Facilities Cost Data 1988*.

Figure 4.10

Installation of Air Handling Unit During Initial Facility Construction

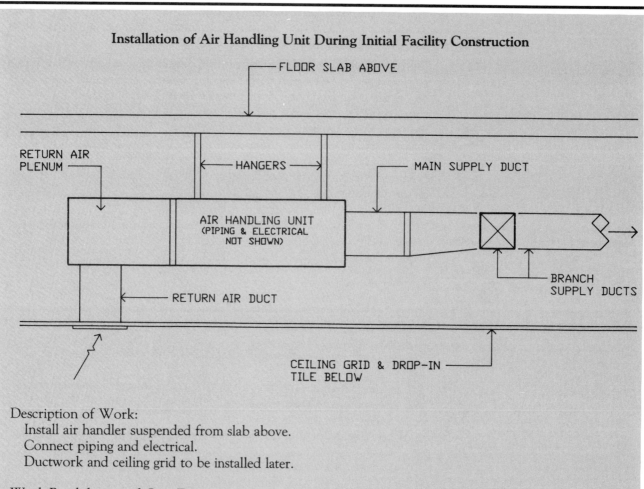

Description of Work:
    Install air handler suspended from slab above.
    Connect piping and electrical.
    Ductwork and ceiling grid to be installed later.

Work Breakdown and Cost Estimate:

Item	Reference*	Material	Labor	Man-Hours
Fan Coil Unit				
3 ton cooling	157-150-0180	$955.00	$ 87	4
Electrical Connection	160-275-0200	$ 2.42	$ 24	1
	Subtotal	$957.42	$111	
	Project Total = $1068.42			

Note: Task is 90% materials, 10% labor.

*Line item from *Means Facilities Cost Data 1988.*

Figure 4.11

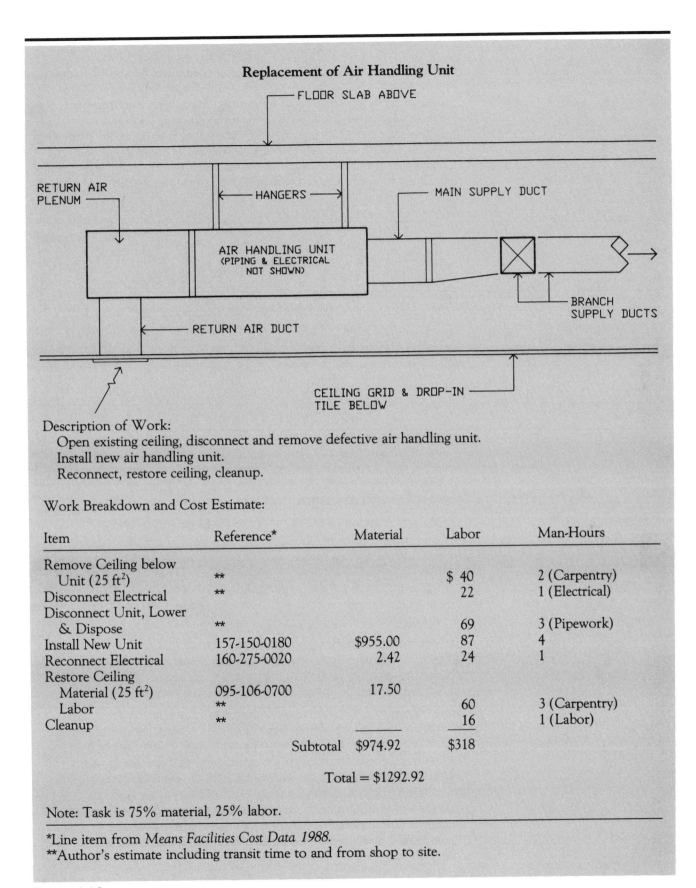

Replacement of Air Handling Unit

FLOOR SLAB ABOVE

RETURN AIR PLENUM

HANGERS

MAIN SUPPLY DUCT

AIR HANDLING UNIT
(PIPING & ELECTRICAL NOT SHOWN)

BRANCH SUPPLY DUCTS

RETURN AIR DUCT

CEILING GRID & DROP-IN TILE BELOW

Description of Work:
 Open existing ceiling, disconnect and remove defective air handling unit.
 Install new air handling unit.
 Reconnect, restore ceiling, cleanup.

Work Breakdown and Cost Estimate:

Item	Reference*	Material	Labor	Man-Hours
Remove Ceiling below Unit (25 ft²)	**		$ 40	2 (Carpentry)
Disconnect Electrical	**		22	1 (Electrical)
Disconnect Unit, Lower & Dispose	**		69	3 (Pipework)
Install New Unit	157-150-0180	$955.00	87	4
Reconnect Electrical	160-275-0020	2.42	24	1
Restore Ceiling Material (25 ft²)	095-106-0700	17.50		
Labor	**		60	3 (Carpentry)
Cleanup	**		16	1 (Labor)
	Subtotal	$974.92	$318	

Total = $1292.92

Note: Task is 75% material, 25% labor.

*Line item from *Means Facilities Cost Data 1988.*
**Author's estimate including transit time to and from shop to site.

Figure 4.12

A comparison of these two examples reveals several points which should be noted. Just as the initial installation of the air handler was a multi-disciplined effort, the replacement project involves work from several shops. During facility construction, the air handling unit was placed prior to the erection of the ceiling. Two visits were required from the electrician: one to disconnect the old unit and again to reconnect the new. During the work, allowance was made for protection of the finished floors, and there was additional cleanup predicted. The replacement project also requires ceiling work twice: once when opening the ceiling for placement of the unit and again to restore the ceiling grid and tiles.

The example above presumes an uncrowded design which minimizes complications during replacement. Within the same example, the design of the building might have included several other problems. Piping for water, sprinkler, or sanitary systems might have been routed under the air handler and would have required removal. Or, if an identically sized unit was not available, the ductwork would have required modification. The development of a reasonable cost estimate requires an extensive survey of the job site to identify any interferences. Accordingly, the necessary craftsman must be identified, the cost estimated, and the work scheduled for the entire project.

Materials, Parts, and Supplies: The cost of major equipment is usually high when compared to the cost of installation. This is not always the case for replacement projects, however. The need to work within an existing building, and the additional labor needed for removal of the old equipment significantly increases the proportion of labor. Nonetheless, the quantity of materials for replacement projects can be easily estimated, and the costs accurately determined in advance. Waste factors for some materials, as noted earlier in this chapter, should be considered.

Tools and Equipment: The tools and equipment needed for replacement projects are the same as those needed for the original installation. Most hand tools used during initial construction are readily available to a maintenance crew. However, the lifting equipment used during the original installation is not routinely part of the maintenance equipment inventory. For this reason, such equipment must be rented or otherwise procured.

Improvement and Modification

The final category of projects executed by the maintenance staff for which cost estimates must be prepared provide either an improvement to the facility or a modification which changes the use of the facility.

Improvement projects are usually identified by the engineering or maintenance staff and carried out voluntarily in anticipation of savings which outweigh the cost of doing the project. The accuracy of the cost estimate is therefore critical to the decision to perform the project. Improvement projects most frequently reduce operating costs by the installation of more efficient equipment or facility components. Installation of additional thermal insulation, more efficient lighting, heating, or air conditioning systems are typical improvement projects. Projects which will reduce maintenance costs are also common.

A modification project is similar to an improvement project, except that the modification project is initiated by the facility user to fulfill a new or unsatisfied existing need. Prior to ordering the modification, the facility user usually requires a cost estimate for the project. Like improvement projects, an accurate cost estimate must be provided to the user so that this individual can make an educated decision. Some modification projects are mandated by changes in the use of the facility. For example, stricter laws governing air pollutants may require the installation of scrubbers in an exhaust chimney. Removal of internal safety hazards, such as asbestos insulation, is another typical modification project.

Productivity: Improvement and modification projects are generally projects which have rarely been previously performed. Therfore, the estimator must depend on published cost guides such as *Means Facilities Cost Data*. However, the estimator should not be completely reliant on these guides. All estimates prepared with cost guide data should be double-checked by the appropriate shop foreman to assure that the figures are reasonable. Since most cost guides are based on cost per unit of work, they are called unit price guides. The unit of measure for each line item varies, being that most likely to relate to the specific activity. For example, painting is estimated by the square foot of wall, carpeting by the square yard, electrical wire by the one hundred foot length. The estimator should be reasonably familiar with the structure of the cost guide and the units usually used for each work area.

When estimating modification projects, the number of each unit required should be accurately determined. This is known as a *quantity takeoff*. Figure 4.13 is an example of a typical modification project for a maintenance organization. Outlined are the scope of work and the salient material specifications. Figure 4.14 is a quantity take-off and estimate for the project described in Figure 4.13.

In Figure 4.14, the quantities are derived from direct measurements from the sketch of the proposed project. They are expressed in units consistent with the units used in *Means Facilities Cost Data, 1988*. The unit prices given are the total of material, labor, and equipment for each line item. The Bare Cost does not include overhead and profit and is used for in-house estimates. If the work is performed by a private contractor, the appropriate figures including overhead and profit should be used. In this example the expected duration of the individual project activities was not determined. If such data was necessary, a separate column would have been shown on the estimate form and the data extracted and calculated from the cost guide line item data.

Materials, Parts, and Supplies: The estimate shown in Figures 4.13 and 4.14 does not specify the exact quantity of material necessary to perform the work. The unit prices used were developed by determining the appropriate type and quantity of material necessary for the unit price, and by pricing such material using average prices. The total material requirement is expressed in dollars. Since improvement or modification projects are not approved without first knowing the estimated costs, the unit price data significantly reduces the amount of time spent on determining the exact material requirements.

If a project is approved, it is necessary to perform a more detailed quantity takeoff, identifying the exact quantity of each type of material used. These identified quantities are then ordered from local suppliers or drawn from existing inventories. This ordering activity is the first measure of the accuracy of the estimate, If the actual cost of the materials varies so much from the estimate that it might affect the original decision to proceed, the maintenance supervisor should be notified and the customer consulted again.

The use of standard cost guides for small modification projects should be approached with caution. Most cost guides are designed for the construction industry. The unit prices in those cost guides imply the use of quantities normal to new construction projects. The advantage of bulk ordering and reduced waste factors makes the unit prices in such publications may be optimistically low compared to the cost of performing a small number of the cited units of work. For example, the cost item in Figure 4.14 shows that painting will cost $.25 per square foot. Taken in the extreme, it can be seen that it is not feasible to perform a project involving the painting of four square feet of wall for the total price of $1.00. The time necessary to travel to the site, open and stir the paint, and then clean up the site and brush after completion is a major cost element. When the size of the painting job grows, these costs are spread over the entire job and become less significant. For these reasons, a common sense examination of the final cost is essential. Where the cost for an item appears excessively low or high, the estimator makes a more detailed estimate, developing the exact number of man-hours and exact quantities of materials and equipment for that task.

Tools and Equipment: The same precautions which apply to materials also apply to the equipment and tools used on modification and improvement projects. The use of the unit prices may not be appropriate for very small tasks. If equipment is rented for a task, the minimum rental period may be in excess of that needed to perform the work. The cost for that minimum rental period is not accounted for by the unit price when that unit price is used for very small jobs. Again, the estimator may need to complete a more detailed quantity takeoff and estimate of equipment requirements.

Modification Project Description
Maintenance Activity: Convert Office to Conference Room

Project Data:

1. Old and New Walls—2 × 4 metal stud with $\frac{5}{8}''$ G.W.B. each side, walls extend through ceiling and are painted.
2. Old and New Doors—3'0" × 7'0" particle core oak face in metal frame. Reuse old door, frame and hardware.
3. Ceiling—2 × 4 suspended ceiling with lay-in tile.
4. Floor Covering—Carpet (not continuous under existing walls).
5. Lighting—Reuse 2 × 4 drop-in fixtures but rewire to new 3-way switches.
6. Heating, Ventilating, Air Conditioning—No change required.

Figure 4.13

Means Forms

COST ANALYSIS

Quantity Take-off & Cost Estimate

SHEET NO.

PROJECT Convert Office to Conference Room

ESTIMATE NO.

ARCHITECT

DATE

TAKE OFF BY: QUANTITIES BY: PRICES BY: EXTENSIONS BY: CHECKED BY:

Item	Qty	Unit	Means Item	Line #	Unit Cost	Extension (nearest Dollar)		
DEMOLITION								
Remove Lighting	8	Ea.	020-708-	2140	12.60	101		
Remove Ceiling	240	SF	020-702-	1580	.35	84		
(Small Job Factor) 50%	240	SF			.18	43		
Remove Walls	180	SF	020-732-	2300	.79	142		
Remove Door	1	Ea	020-706-	0200	8.30	8		
Remove Frame	1	Ea	020-706-	2000	22.75	23		
Remove Carpet	240	SF	020-712-	0400	.13	31		
Construct New Walls								
Framing & G.W.B	256	SF	092-620-	3800	1.85	474		
Door Frame (new)	1	Ea	081-118-	0100	79.85	80		
Door Frame (Reinstall)	1	Ea	081-118-	0100	21.00	21		
Door (new)	1	Ea	082-062-	2180	94.05	94		
Door (Reinstall)	1	Ea	082-062-	2180	27.05	27		
Door Hardware								
Hinge	1	Pr	087-116-	0400	40.00	40		
Lockset	1	Ea	087-120-	0100	44.15	44		
Kick Plate	1	Ea	087-118-	0010	24.90	25		
Painting (Walls)	512	SF	099-224-	0800	.25	128		
Varnish Door	1	Ea	099-216-	3000	48.95	49		
Carpet (Nylon)	27	SY	096-852-	3200	16.93	457		
Carpet Pad	27	SY	096-852-	9200	3.61	97		
Ceiling (2x4 Tile)	240	SF	095-106-	0810	1.45	348		
Replace Lights (labor)	8	Ea	166-130-	0600	40.00	320		
Wire Switch	2	Ea	162-320-	0800	15.61	31		
Wire Lighting	100	LF	161-145-	0250	110.00	110		
						2777		

Figure 4.14

Summary

The cost of performing maintenance activities consists of the individual costs for the following elements: labor, materials, supplies, spare parts, equipment, and overhead support expended. Each of these components has its own individual characteristics peculiar to maintenance activities which must be understood if costs are to be reasonably controlled.

A first step in controlling the cost of maintenance is to establish realistic cost estimates of individual maintenance activities. These estimates are used to determine whether projects should be undertaken. The magnitude of the estimated labor, material, and equipment costs can be used to determine the area where strict management supervision may reap the greatest benefits in reducing overall costs. The estimates also become the yardstick by which progress is measured. For each different type of direct maintenance work, the proportions of labor, material, and equipment may vary. Different techniques are used in developing the cost estimate for each type. A consistent method for estimating produces consistently accurate estimates. This consistency is a major contributor to full control of maintenance costs.

Chapter Five

COST-BASED MAINTENANCE DECISIONS

Chapter Five

COST-BASED MAINTENANCE DECISIONS

The objective of any maintenance program is to provide a completely usable facility at minimal cost. In order to accomplish this, costs should be estimated in advance. In the previous chapter, methods were presented for producing reasonably accurate cost estimates for various maintenance activities. In making these decisions, the maintenance manager uses those cost estimates to make decisions which affect the state of maintenance of the facility and the overall maintenance budget. The manager is concerned with both the short- and long-term costs of maintenance, and, therefore, any expenditure decision for maintenance should be carefully considered. Short-term costs are the exact costs of the part and its associated installation costs. Long-term costs include maintenance needs, operating costs, and the overall effect the part has on the life of the building.

Each maintenance activity is an investment into the overall life of the building or facility. (See Chapter 1 for a description of the life cycle of a facility.) General maintenance activities are investments to prevent further degradation and more costly repairs. Preventive maintenance is an investment to prevent short-term failure and extend the useful life of the system or component. (See Chapter 11 for a more detailed description of preventive maintenance.) An improvement project reaps long-term benefits, such as reducing operating costs and reducing the frequency of future maintenance needs. If maintenance activities are viewed as investments, then the decision to execute those projects should be analyzed as a business would analyze a financial investment. Options should be examined and compared. Risks should be analyzed. Paybacks for maintenance projects should be determined.

Making maintenance decisions involves the comparison of two or more alternative maintenance actions. The best option is that which provides the most benefits for the lowest cost. The benefits are measured by the manner in which the decision allows the proper continued use of the facility. The lowest cost is a total of short- and long-term costs. The short-term costs are the costs of executing a maintenance project. Long-term costs are the estimated costs of operating and maintaining the results of that maintenance project.

Two methods are described in this chapter for comparing alternative maintenance actions: the *present value method* and the *annual cost method*. Each is discussed in the following sections.

Present Value Method

If a business expects to realize a profit on its investments, a decision is made concerning when and where to invest the capital of the business. Maintenance activities are investments and the timing of these investments is critical; thus, when comparing the costs of the two alternatives, it is necessary to consider the *apparent present value* of both the short-term and long-term costs.

In order to compare present and future costs of maintenance, these costs must be compared at a common point in time. Present value analysis expresses future costs in today's dollars, or "how much money should be invested today at a particular discount factor to appreciate sufficiently to meet that future cash flow?" Through proper calculations, the present value of any future investment can be determined and a total of several present and future costs can be expressed as a single present time investment.

The present value method uses a *discount factor* for each cost item in order to establish what the actual value of a maintenance project will be, considering both the present and the future. The discount factor is the percentage of real growth that a business expects to earn on any investment. This real growth goal is measured after considering the effects of inflation on the value of the dollar. If a business expects to earn ten percent profit on an investment (a maintenance project), then its discount factor is ten percent. For example, consider a maintenance project which will cost $1000 if executed today or the same $1000 if executed one year later. The two costs appear equal. But if only $909.09 was invested and earned a ten percent profit in one year, it would be worth $1000.00 ($909.09 + 10%) in one year. Therefore the present value of $1000.00 one year in the future is $909.09 in today's dollars. The options are to invest $1000.00 today and execute the project immediately, or to apply $909.09 elsewhere in the business where it will grow into $1000 and execute the project one year from today. Obviously the project would be delayed one year with a $91 apparent savings.

The discount rate should not be confused with the inflation rate of the cost of goods and services. Normal inflation causes a yearly increase in the cost or value of most maintenance activities. However, a business still wants to achieve a real ten percent profit on any investment (or maintenance project) above and beyond inflation. To exactly account for inflation, future costs would be adjusted upward to reflect inflation, the discount rate increased (to above ten percent) to provide the net ten percent profit desired, and the present value calculated. Since the result of this exact analysis is nearly identical to simply disregarding the inflation and using the desired ten percent discount rate, the effect of inflation is generally disregarded.

The use of ten percent as an example is not accidental. Realizing a true ten percent growth on an investment is a reasonable business goal for most maintenance organizations. When a maintenance organization is part of a profit-making business, the corporation's desired discount rate should be substituted. Even if a maintenance organization is part of a non-profit public agency, the discounting method should be used as if the supporting tax dollars or contributions were being invested in the maintenance work.

Single Time Cost Items

The use of the present value theory of comparing costs relies on the identification of cost expenditures at the present and in the future. The present value of a cash expenditure made at the time of the analysis represents the full value of the expenditure. The present value of future expenditures is less than the full value, the amount depending on the discount rate and the length of time between the present and the time of that future expenditure.

For the purpose of describing the process and elements of present value analysis, the following standard abbreviations are used:

P = a single cash flow at the present time

F = a single cash flow at a future time

A = a regularly recurring cash flow or annuity

P/F = the present value of a single future cash flow

F/P = the future value of a single present time cash flow

P/A = the present value of a series of regularly recurring future cash flows

A/P = a uniform cash flow equal to a single cash flow at the present time

A/F = a uniform cash flow equal to a single cash flow at some time in the future

i = the discount rate or percentage for a particular period (expressed as a decimal)

n = the number of equal length periods of time

The annuity, or "A" cited above, represents any series of equal cash flows occurring at regularly spaced intervals, such as the uniform payments on a loan. The discount rate, "i," is the percentage associated with the period denoted by "n" (the number of equal-length periods of time). Examples throughout this chapter deal with annuities occurring at annual intervals, thus "n" will always represent the number of years an annuity is received. The discount or interest rate ("i") is the annual interest rate, and in this chapter the rate of ten percent is used in all examples.

In the most simple cases, the calculation of the future value "F" (a single cash flow at a future date) of a present time cash flow "P" (a single cash flow) is made using the following formula:

$$F = P + (P \times i)$$

This formula can be simplified to read:

$$F = P \times (1 + i)$$

Therefore, the future value (one year from present) of a present cash flow of $100 at a discount rate of ten percent is calculated as follows:

$$F = \$100 \times (1 + 10) = \$110$$

If the future value was desired for two years in the future, the annual discount rate would be applied twice, as shown below.

$$F = P \times (1 + i) \times (1 + i) \quad \text{or} \quad F = P \times (1 + i)^2$$

Continuing this process, the future value, "F," of a present cash flow, "P," at "n" years in the future is calculated as shown below.

$$F/P = P \times (1 + i)^n$$

The term "$(1 + i)^n$," can be calculated for any interest rate or number of periods and is called the "Compound Amount Factor."

Conversely, to calculate the present value, "P," of a future cash flow, "F," the formula is reduced as follows:

$$P/F = F \times \left(\frac{1}{(1 + i)^n} \right)$$

The term multiplied by "F" in this case is called the *present value factor* and can be calculated for any "n" and any "i," and can be multiplied by any future cash flow to compute the present value of that cash flow. A *time line* such as that shown in Figure 5.1, can be used to visually describe the cash flows for which the present value is being calculated.

The time line in Figure 5.1 shows three separate cash flows. A $2000 cash expenditure is shown at the end of year 2. A $1000 cash expenditure is shown at the end of year 4, and a $3000 cash income is shown at the end of year 5.

A table of Present and Future Value Factors is presented in Figure 5.2. This table can be used to calculate the three cash flows in Figure 5.1.

This table contains the various factors used in present value calculations, and represents the factor values for an interest rate of ten percent. To use the table to compute the present value of a future cash flow (P/F), the factor P/F is selected for the appropriate period "n" and multiplied by the value of that future cash flow. For clarity, the following standard format will be used for describing these calculations:

P = F(P/F, n, i)

Figure 5.3 shows the calculation of present value for the three cash flows in Figure 5.1.

In the examples above, the two cash expenditures were considered as positive values and the one cash income was given a negative value. The convention for positive/negative is optional. Since the comparisons made by maintenance managers deal primarily with maintenance costs, all examples in this book assume that cash expenditures have positive values and cash incomes have negative values.

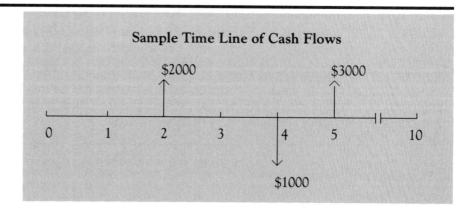

Sample Time Line of Cash Flows

Figure 5.1

Present and Future Value Factors for 10% Discount Rate

	Single Cash Flows		Uniform Recurring Cash Flows		
n	P/F	F/P	P/A	A/P	A/F
1	0.909091	1.100000	0.909091	1.100000	1.000000
2	0.826446	1.210000	1.735537	0.576190	0.476190
3	0.751315	1.331000	2.486852	0.402115	0.302115
4	0.683013	1.464100	3.169866	0.315471	0.215471
5	0.620921	1.610510	3.790787	0.263797	0.163797
6	0.564474	1.771561	4.355261	0.229607	0.129607
7	0.513158	1.948717	4.868420	0.205405	0.105405
8	0.466507	2.143589	5.334927	0.187444	0.087444
9	0.424098	2.357948	5.759024	0.173641	0.073641
10	0.385543	2.593743	6.144568	0.162745	0.062745
11	0.350494	2.853117	6.495602	0.153963	0.053963
12	0.318631	3.138429	8.813693	0.146763	0.046763
13	0.289664	3.452272	7.103357	0.140779	0.040779
14	0.263331	3.797499	7.366688	0.135746	0.035746
15	0.239392	4.177249	7.606081	0.131474	0.031474
16	0.217629	4.594975	7.823709	0.127817	0.027817
17	0.197845	5.054472	8.021554	0.124664	0.024664
18	0.179859	5.559919	8.201412	0.121930	0.021930
19	0.163508	6.115911	8.364921	0.119547	0.019547
20	0.148644	6.727503	8.513564	0.117460	0.017460
21	0.135131	7.400253	8.648695	0.115624	0.015624
22	0.122846	8.140279	8.771541	0.114005	0.014005
23	0.111678	8.954307	8.883219	0.112572	0.012572
24	0.101526	9.849738	8.984744	0.111300	0.011300
25	0.092296	10.834712	9.077041	0.110168	0.010168
26	0.083905	11.918183	9.160946	0.109159	0.009159
27	0.076278	13.110002	9.237223	0.108258	0.008258
28	0.069343	14.421002	9.306566	0.107451	0.007451
29	0.063039	15.863103	9.369606	0.106728	0.006728
30	0.057309	17.449413	9.426915	0.106079	0.006079

Figure 5.2

Recurring Costs

The annual preventive maintenance activities associated with any piece of equipment generate recurring costs over the life of the equipment. This is a common occurrence in a maintenance program. The annual cost of operating a piece of equipment, or even an entire facility, is expressed as a *constant annual value*. When a series of equally valued cash flows occur at annual intervals, a series of individual calculations of present value could be made, one calculation for each annual cash flow. However, this is tedious and unnecessary. When the cash flows are equal and occur at annual intervals, the calculation of the present value of that series of cash flows can be performed using a factor from Figure 5.2. This is illustrated in Figure 5.4, which shows the calculation of the present value of a series of $100 expenditures at yearly intervals for six years. In this example, the present value of a six-year series of cash expenditures of $100 at a ten percent discount rate is shown to be equal to $435.53. This value is the amount of money which, if invested at a rate of ten percent, would be sufficient to meet the individual $100 cash flow occurring at the end of each of six years. Figure 5.5 demonstrates this by tracing the cash balance of such an account as it increases by ten percent each year and is then reduced by $100 each year.

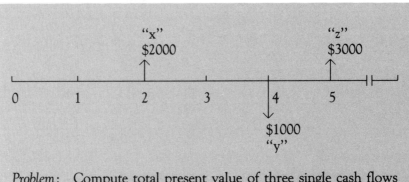

Problem: Compute total present value of three single cash flows described on the above time line.

Solution: (Using factors from Figure 5.2)

P of "x" = $2000 × (P/F, n = 2, 10%)
 = $2000 × (.826446)
 = $1652.89

P of "y" = −$1000 × (P/F, n = 4, 10%)
 = −$1000 × (.683013)
 = −$683.01

P of "z" = $3000 × (P/F, n = 5, 10%)
 = $3000 × (.620921)
 = $1862.76

Total Present Value = $1652.89 − $683.01 + $1862.76
 = $2832.64

Figure 5.3

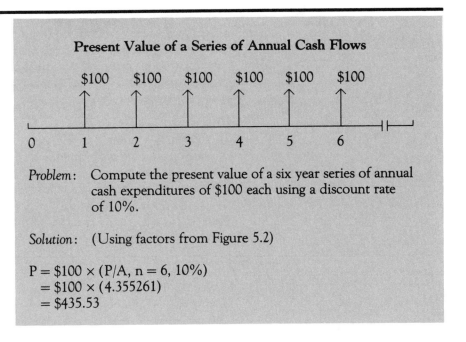

Present Value of a Series of Annual Cash Flows

$$\$100 \quad \$100 \quad \$100 \quad \$100 \quad \$100 \quad \$100$$

0   1   2   3   4   5   6

Problem: Compute the present value of a six year series of annual cash expenditures of $100 each using a discount rate of 10%.

Solution: (Using factors from Figure 5.2)

$$P = \$100 \times (P/A, n = 6, 10\%)$$
$$= \$100 \times (4.355261)$$
$$= \$435.53$$

Figure 5.4

Example of Cash Flows and Present Value

Given: Present value of a series of six annual cash flows at 10% is $435.53

Example:

Starting Balance	435.53
+10%	43.55
−$100	−100.00
End of Year 1	379.08
+10%	37.91
−$100	−100.00
End of Year 2	316.99
+10%	31.70
−$100	−100.00
End of Year 3	248.69
+10%	24.87
−$100	−100.00
End of Year 4	173.56
+10%	17.35
−$100	100.00
End of Year 5	90.91
+10%	9.09
−$100	−100.00
	00.00

Figure 5.5

The two types of payments, a single time payment and a series of equal recurring payments, can be combined to describe the present value of any piece of equipment. Figure 5.6 is an example of the total present value cost of owning, operating, and maintaining a 50KW emergency generator set within a facility. The dollar costs shown are derived from *Means Facilities Cost Data, 1988*.

Restrictions

The methods presented in this book are strictly rudimentary methods of comparing alternatives. When dealing with such factors as varying annual costs, varying inflation rates for materials or services, cost benefit ratios, and predicted rates of return on investments, the reader should consult an advanced text in the field of economic analysis.

The present value calculation is of little value unless used in the comparison of alternatives. An important consideration, then, for the use of present value analysis is the service life of the various alternatives; the life of each alternative must be equal for the study to be valid. Since studies often deal with varying equipment lives, it is necessary to seek a *lowest common multiple* for the alternatives. If, for example, two alternatives have predicted service lives of two and three years respectively, it is necessary to repeat each set of cash flows to reach a

Present Value of the Ownership, Operating, and Maintenance Costs of an Emergency Generator

Given: A 50 KW diesel-powered generator set including fuel tank and transfer switch with estimated 15 year life

Initial Costs (including installation): $20,500

Operating Costs: (100 hours per year assumed)
   Fuel: 10 gal/hr @ $1.10/gal × 100 hrs = $1100 annually

Maintenance Costs:
   Overhaul engine at years 5 & 10   = $4000
   Test run monthly   = $300 annual
   Oil & filter change semiannual   = $270 annual

Summary:
   Single Event Costs
      Installation $20,000 at year 0
      Overhaul $4,000 at year 5
      Overhaul $4,000 at year 10

   Annual Recurring Costs
      Fuel $1100
      Testing   300
      Oil & filters 270
                    ─────
                    $1670 annually

Figure 5.6

common end point of the comparison. In this case, a life of six years means that the three-year alternative is repeated twice and the two-year item three times. Figure 5.7 shows the time lines for such a comparison.

The service lives illustrated are matched up in relatively few repetitions. However, when service lives are not easily matched (such as 7 and 9), the analysis can become unwieldy. In such cases, the uniform annual cost method should be used.

Uniform Annual Cost Method

The uniform annual cost method is used primarily to compare alternatives with different service lives. While the present value method brought all future costs back to their equivalent value at the present time, the uniform annual cost method expresses all present and future costs as equivalent recurring cash flows.

Comparison of Alternatives with Unequal Service Lives

Given:

	Alternative A	Alternative B
Purchase Price	$10,000	$14,000
Operating Costs	$2,000	$1,000
Service Life	2	3

Time Lines

Alternative A

Alternative B

Figure 5.7

117

Two alternatives can be compared by simply dividing the total present cost by the service life to receive an annual cost equivalent. This, however, would not address the value of money with respect to time. The uniform annual cost method accommodates the discount rate or desired return on investment. The annual value computed for a present time investment should represent the payment required to retire a loan in the amount of the investment.

Single Time Cost Items

Single present time item costs are handled simply in the uniform annual cost method. The cost is converted into an equivalent annual payment based on the service life of the alternative. The computed values of this A/P factor (a uniform cash flow equal to a single cash flow at the present time) are shown in Figure 5.2. The uniform annual cost is equal to the cash flow multiplied by the A/P factor in the table where "n" equals the service life. If there is an anticipated intermediate single cash flow during the service life (such as an overhaul or major maintenance action), the value of that cash flow is brought back to the present using present value method calculations. That present value is added to the initial cash flow, and the total is then converted to a uniform annual cost.

Recurring Costs

Uniform recurring costs are treated as a cash flow without respect to the discount rate or the service life. They are simply added to the uniform annual cost for the single cash flows. The resulting total is the uniform annual cost used for comparison purposes. Figure 5.8 shows the uniform annual cost comparison for alternatives with different service lives (from Figure 5.7).

Repair Versus Replacement Analysis

One of the most common decisions facing a maintenance manager is the choice between repairing or replacing a defective piece of equipment. This decision should be made considering all the costs and impacts of each alternative, and also considering the alternatives as investments. The principles of the present value and uniform annual cost methods should be used.

There are always individual characteristics associated with each alternative: partial repair, full overhaul, replacement in kind, or replacement with different equipment. One choice may be immediately available, another may offer reductions in future maintenance costs, while still another might offer the opportunity for increased standardization. All of these varying assets must be reduced to a simple cost value in order to allow comparison of the alternatives.

Failure and Maintenance Rate

When a piece of equipment has failed and requires either repair or replacement, the results of a repair may not produce the same continuing reliability as that of a new replacement unit; a repair does not usually restore the unit to a "like new" condition. Internal portions of the equipment which did not fail at this time may be excessively worn and may produce another near-term failure.

A major failure of a piece of equipment may occur for several reasons: incomplete or improper maintenance; the end of its predicted service life; or a defect in manufacture or installation. When comparing the alternatives of repair versus replacement, the facilities manager must determine how long the repaired equipment will last before replacement is inevitable. If the failure occurs early in the service life of a component or system, it is related most likely to a defect in manufacture, and further failures would be expected to be similar in extent and frequency. For such failures, the predicted service life should not be altered. When a failure occurs nearer the end of the predicted service life, that failure generally indicates a slightly shorter service life.

Regardless of the indications given by the failure, the maintenance manager should be aware of predicted major maintenance activities associated with the equipment. These activities, which may include a major rebuilding or overhaul at mid-life or the replacement of several wearing parts, must be addressed in the cost comparison. Data from the equipment manufacturer and similar data collected from the facility's historical maintenance records can be used to predict the frequency and cost of future planned maintenance. Since the failed equipment might not be replaced with an identical unit, the operating costs of the existing and the proposed equipment must be compared.

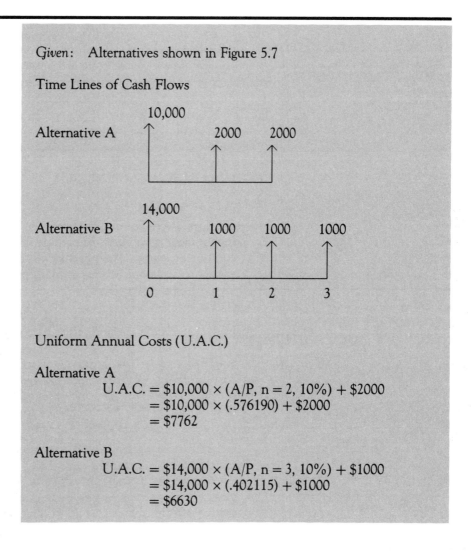

Given: Alternatives shown in Figure 5.7

Time Lines of Cash Flows

Uniform Annual Costs (U.A.C.)

Alternative A
$$U.A.C. = \$10,000 \times (A/P, n = 2, 10\%) + \$2000$$
$$= \$10,000 \times (.576190) + \$2000$$
$$= \$7762$$

Alternative B
$$U.A.C. = \$14,000 \times (A/P, n = 3, 10\%) + \$1000$$
$$= \$14,000 \times (.402115) + \$1000$$
$$= \$6630$$

Figure 5.8

Loss of Facility Use

The impact on facility use is a major factor influencing the decision to repair or replace a system or component, because any delay to equipment or system restoration is costly. When the failure prohibits facility use, the cost can be substantial. Lost production means lost profits or additional overtime after the repair to compensate for the lost time. Cancelled operations may lead to financial damages due to failure to meet contractual requirements, such as when a convention or meeting is cancelled due to facility problems, for example.

Some costs are difficult to assess in exact financial terms. The lack of air conditioning in an office may not preclude use of the office, but is likely to lead to some loss in productivity for the workers in that space. The costs become firmly defined if this loss in productivity leads to the need for overtime to catch up. Other failures do not encumber the continued use of a facility in financial terms. Failure of a fire alarm system may not be noticed by the employees, but places the building occupants and contents at a greatly increased risk. This kind of risk is difficult to quantify, but often leads to a decision to take whichever action results in the fastest restoration of system operation, regardless of cost. When a failed component or system is backed up by a similar system, the cost of delays is less significant. The risk of total loss of facility use is increased, and some redundant systems may have some increased operating cost. For example, it is usually more expensive to run an emergency generator for building power than to use electric power from the local utility.

In assessing the cost of loss of equipment function, the reasons for the delay in system restoration must be determined. If the replacement equipment and work crew are not immediately available, the source of supply and time for normal or emergency shipping must be identified. If the replacement is not immediately available, repair is generally faster, and thus less costly, unless a work crew is not available to perform the repair. All of these factors lead to many different options beyond the two simple choices of repair or replacement; there are many variations of each one. The example in Figure 5.9 is a simple comparison of the repair or replacement of a failed air conditioning condenser. The costs and the resulting conclusion used in the example are estimated and are for demonstration purposes only; this comparison should not be considered a definite indication for action in a similar failure. Since the replacement equipment has an identical service life to that of the in-place equipment, but that service life will be staggered, the use of the present value method comparison is not applicable. Uniform annual cost methods should be used.

Labor-Saving Devices

In the search for methods to reduce maintenance costs, much emphasis is placed on reducing labor costs. Since the salaries or wages paid to employees are relatively stable, concentration in the area of reducing the cost of labor is placed on finding devices which will improve worker productivity. If sufficient increases in productivity can be achieved, the overall staffing may be reduced or the existing staff may be able to accept increased workloads associated with an expansion of facilities.

Generally the availability of labor-saving devices is made known to the maintenance manager through direct contact with salesmen and through advertisements in trade literature. Individual workers might also bring ideas from previous employment or of their own invention. There is no single right time to choose to purchase a labor-saving device. However,

Repair versus Replacement Example

Situation: An air conditioning condenser has failed near the end of its useful life. Repairs can be made which will delay replacement for two years. Repairs can be completed in 4 days while 6 days are required to acquire and install a new condenser.

Options:

I Repair the failed condenser immediately and schedule replacement for two years hence. Scheduled replacement will not interrupt facility operation.

II Replace condenser immediately. Experience loss in facility use for 6 days.

Estimates:

Repair cost: $3000
Replacement cost: $10,500
Operating costs: Equal for new or repaired condenser
Loss of facility use: $500 per day
Service life of new unit: 15 years

Method of Comparison:

Since the repaired unit will be replaced in two years, option I and II will never be synchronized. Uniform annual cost methods must be used.

Analysis:

Option I (17 year life)

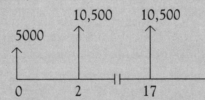

1. Convert replacement at year 2 to present value
   $P_{(10,500)}$ = $10,500 × (P/F, n = 2, 10%)
   = $10,500 (.826446)
   = $8678

2. Convert total of year 0 and present value of replacement to uniform annual cost
   P = $5000 + $8678 = $13678
   A = $13678 × (A/P, n = 17, 10%)
   = $13678 × (.124664) = $1705

Option II

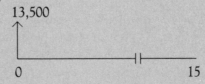

A = $13,500 × (A/P, n = 15, 10%)
= $13,500 × (.131474) = $1775

Figure 5.9

the scheduled replacement of a piece of maintenance equipment is a logical time to examine the advantages of varied replacement options in terms of reducing the labor required.

Changes in Productivity

The productive output of the daily tasks of employees can be altered in most cases. The carpenter may benefit from the use of certain power tools, the air conditioning mechanic may diagnose problems faster with enhanced leak detection equipment, the janitor may be able to clean a greater area in less time with powered cleaning equipment. Most advertisements for labor-saving devices emphasize an increase in productivity. Such claims should always be examined for validity. The improvement in productivity of some tools or equipment can be determined using production rates published in cost guides such as *Means Facilities Cost Data*. Individual line items state the expected hourly or daily production rates of various equipment engaged in construction and maintenance tasks. In many cases, the labor-saving device under consideration may already be in limited use within the maintenance organization, and the attendant improvements in productivity can be determined from historical records and direct comparison.

Like any investment of funds previously described in this chapter, a direct cost analysis should be performed in order to evaluate the purchase and use of a labor-saving device. The increased cost of the device should be weighed against the reduced labor costs for its operation. In addition, the value of the existing equipment must be estimated, since any value that can be gained through the sale of the equipment, or any tax advantage which might result from its donation to others, should be properly applied in the cost analysis. If not saleable, the cost of disposal of failed equipment and the cost of any additional training must also be included in the cost comparison. There may be significant resistance to the imposition of labor-saving devices if their use could lead to staffing reductions. Such resistance may be an insurmountable factor. Therefore, the feasibility of obtaining the desired increase in production must be thoroughly examined by the foreman and maintenance manager before the expected productivity rate is set for the various options in the cost study. Figure 5.10 is a cost comparison of two different purchase options for lawn mowing equipment. It represents a typical case comparing an inexpensive, but labor intensive piece of equipment versus a more costly, but more productive option.

Communicating the Decision

Once the decision to purchase labor-saving equipment or tools is made, the predicted increase in productivity must become a reality. Often the workers' motives desiring a new tool or more productive equipment are distinctly different from the management goal of reducing labor costs: the worker sees the tool as a means of making his or her job easier. That the tool will enhance the worker's productivity is not likely to be a major consideration for the worker. It is important, therefore, that the maintenance manager follow up the purchase of a labor-saving device with firm communication of the expectations associated with the equipment purchase. If this is not the case, a worker will generally complete a certain task in the same period of time as it took without the labor-saving device. For this reason, the maintenance manager and shop foreman should monitor the use of the new equipment to ensure that the expected productivity is achieved.

Comparison of Alternatives with Differing Productivity Outputs

Situation: A facility containing 6 acres of lawn performs its own lawn mowing. The current 40″ riding mower is near the end of its useful life. A replacement will be made. Lawns are mowed 20 times annually.

Problem: Should the mower be replaced in kind or should a mower with higher productive capacity be purchased?

Description of Alternatives:

Option I   40″ mower, capable of mowing 300,000 square feet per day, expected life 4 years.

Option II   Small tractor and 3-reel gang mower capable of mowing 930,000 square feet per day, expected service life 8 years.

Estimated Costs:

Option I   Purchase Price $2300
Operating Costs:
   Operator—common laborer @ $16.55/hr
   6 acres mowed at 300 m.s.f/day
      = 6.97 hours per mowing
      = $115.35 per mowing
      = $2307 per year
   Fuel = $90/hr = $125.46 per year
   Overhaul at midlife = $1000 at year 2

Option II   Purchase Price $7000
Operating Costs:
   Operator—common laborer @ $16.55/hour
   6 acres at 930 m.s.f. per day
      = 2.25 hours per mowing
      = $37.24 per mowing
      = $744.80 per year
   Fuel = $6.00 per hour
      = $270 per year
   Overhaul at midlife = $3000 at year 4
   Salvage value at year 8 = $1500

Present Value Analysis:
   Analysis based on an eight year cycle.

Option I

Figure 5.10

Additional training is often necessary to provide safe operation of the new equipment at its desired productive output. The training should be readily provided to ensure that workers achieve the desired increase in production.

Evaluating Reimbursable Modifications

The reasons for initiating a modification project is not much different from those which lead to the purchase of a labor-saving device or the decision of repair versus replacement. In requesting a facility modification, the facility user or customer has made a determination that the cost of the modification project is justified by the services or benefits it will provide. When such projects promise reimbursement through increased benefits, however, the maintenance manager should thoroughly identify the total costs of performing the modification project. Previous examples in this chapter have dealt with the mechanics of conducting a formal cost comparison between various alternatives. The same cost comparisons should be made when considering several different methods for accomplishing the modification.

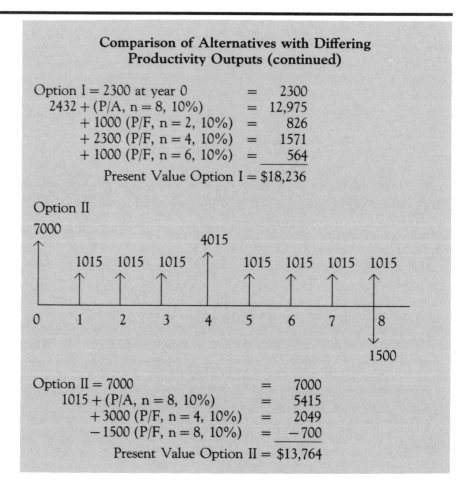

Comparison of Alternatives with Differing Productivity Outputs (continued)

Option I = 2300 at year 0 = 2300
2432 + (P/A, n = 8, 10%) = 12,975
+ 1000 (P/F, n = 2, 10%) = 826
+ 2300 (P/F, n = 4, 10%) = 1571
+ 1000 (P/F, n = 6, 10%) = 564

Present Value Option I = $18,236

Option II

Option II = 7000 = 7000
1015 + (P/A, n = 8, 10%) = 5415
+ 3000 (P/F, n = 4, 10%) = 2049
− 1500 (P/F, n = 8, 10%) = −700

Present Value Option II = $13,764

Figure 5.10 (continued)

Defining the Extent of Reimbursement

If a maintenance organization spends most of its time on routine and preventive maintenance, the cost accounting system for that work is generally not sophisticated enough to provide an exact cost for each maintenance activity. Controlling certain types of maintenance costs for such organizations does not produce enough savings to justify the expense that collecting and monitoring those costs would incur. In such cases, maintenance activities are carried as one general line item in the budget and are not attributed to individual tasks.

When the need arises to charge a customer for a modification project, it will be necessary to identify the *full cost* of such a project. The simple cost of materials, labor, and equipment used in the project generally significantly underestimates the true cost. There are many hidden costs to consider. Figure 5.11 is a listing of hidden costs and the recommended means for determining the proper value to place on these items.

Evaluating Facility Improvement Projects

As defined in Chapter 1, an improvement project is one which is accomplished with the direct intent of reducing overall costs or improving profits. In other words, an improvement project should stand on its own merits, and the benefits should unquestionably justify the expenditure of the funds. However, this is only true if the decision follows a strict analysis and comparison of costs.

Definition of Benefits

An improvement project is intended to improve the financial state of the facility or its maintenance program. There are several benefits to improvement projects. An improvement project may increase profits through expanded facilities and business opportunities. The purchase of a labor-saving device is an example of an improvement project utilized to reduce labor costs. The rapidly increasing costs of energy and water have given rise to a variety of products and methods designed to reduce utility consumption. For this reason, many improvement projects are aimed at reducing the operating costs of the facility, particularly in this area. The benefits of this type of improvement project are predictable; they will reduce consumption by a known amount. Several common energy saving projects are listed below.

- Installation of flow-restricting faucets or shower heads which will reduce hot water consumption.
- Installation of flow restrictors in toilets to reduce water consumption.
- Installation of energy-saver ballasts in fluorescent lamps.
- Delamping light fixtures, reducing lighting levels to the minimum acceptable level.
- Installation of additional thermal insulation in building exteriors to reduce heating or cooling costs.
- Replacement of boilers or furnaces with more efficient units.
- Installation of a shallow well for lawn watering reducing reliance upon commercial water supplies.
- Installations of night setback thermostats.

Simple Pay-backs

The easiest economic comparison for evaluating improvement projects is the *simple pay-back*. This calculation is the result produced by dividing the cost of the improvement by the predicted annual savings, thereby calculating the number of years of operation of a device which will produce savings equal to the cost of the project. This formula is shown below:

$$\frac{\text{cost of improvement}}{\text{predicted annual savings}} = \begin{array}{l}\text{number of years of operation to produce}\\\text{savings equal to cost of project}\end{array}$$

The facilities manager should be careful to include all costs associated with the improvement. Advertisements for devices which will produce savings often predict a pay-back based on the predicted savings and the purchase price of the device. However, an attractive pay-back may not actually result when the cost of installation is added to the purchase price.

Hidden Costs for Modification Projects

Indirect Activities Supporting Project	Recommended Method for Determining Cost
Shop Foreman Direct Supervision	Actual Cost (1)
Shop Foreman Overhead	Shop OVHD Rate (2)
Shop Tool and Equipment Expense	Shop OVHD Rate (2)
Management Supervision	General Overhead % (3)
Management Estimating	General Overhead % (3)
Purchasing, Inventory Control	General Overhead % (3)
Computerized Work Tracking	General Overhead % (3)
Payroll Preparation	General Overhead % (3)
Cost Accounting	General Overhead % (3)
Compliance Inspections	Actual Cost (1)
Engineering	Actual Cost (3)
Profit	Percentage Added

Notes
(1) Actual hourly wage rate, including overhead multiplied by number of hours expended.
(2) Pro rata share of these expenses added as an hourly cost or as percentage.
(3) These costs are normally added as a percentage of actual direct labor and materials. This percentage rate is determined by comparing the total annual cost of all overhead activities with the total annual direct maintenance costs.

Figure 5.11

126

There is no set pay-back period which should be used to validate improvement projects. Generally, a project with a pay-back of less than one year should definitely be executed immediately, because the funds necessary for the project are already available within the yearly budget and can be reassigned to execute the project. Any project with a pay-back greater than one year requires an increase in the budget. For such cases, most organizations establish a policy that all projects with pay-backs less than a certain number of years are scheduled immediately (if funding can be made available). When pay-backs are greater than three years, it may be advantageous to perform a present value cost comparison to determine an exact pay-back evaluation. When improvement projects are expected to produce increased sales or production, a formal present value comparison should definitely be made. Figure 5.12 is a sample calculation of the simple pay-back for the installation of flow-restricting shower heads. These shower heads use less water from a municipal water system and thus reduce costs for heating the hot water.

Quantifying Subjective Analysis

To this point the various decisions described in this chapter have all dealt with identifiable project costs, benefits, and savings. The analysis, in each case, was an exact comparison of two or more alternatives. In many cases, however, the maintenance manager must make decisions based on inexact or subjective data, which may, in turn, produce benefits that cannot be assigned a specific dollar value. For example, one of the maintenance manager's goals may be courteous treatment of the customers. The benefit expected from this minor effort is an improved relationship with the customer which may result in better cooperation. This benefit is difficult to quantify but has some perceived value. However, such a goal requires some expenditure of managerial time (e.g., overseeing the workers to ensure courteous treatment of customers).

Another typical area where value is difficult to quantify is the appearance of a facility. Different wall coverings create atmospheres which are "rich and classy," "tasteful and professional," or "austere." Each one has a different maintenance cost and benefit. The facility manager or owner, working with the maintenance manager, must form a policy for the appearance of the facility interior or exterior. In this case, no exact benefit can be defined, but the decision maker may rely on personal preference or "gut reaction." This section contains a description of the *weighted average* method of analyzing the value of subjective decisions.

Setting Evaluation Parameters

When a choice is made between two alternatives of known cost but of a single unknown benefit, the decision maker can rely on personal judgment of the relative value of each. Is the extra cost justified by the better appearance? When, however, the benefits can be classified in more than one category, and each alternative satisfies each category differently, the decision is not straightforward. For example, a new car buyer contemplating several models considers the price, reliability, styling, functionality, acceleration, and handling of the various choices. Each of these characteristics has a different relative value to the car buyer, and each car model measures up differently in each category. The buyer weighs the definitive price, acceleration, and reliability, as well as the subjective elements of styling and handling.

In using the weighted average method of evaluating alternatives, the maintenance manager identifies the characteristics of a project that influence the decision between the alternatives. Characteristics are both objective, such as initial cost and predicted maintenance cost, and subjective, such as appearance and reliability. Some alternatives offer features that are not essential but provide some improved, perceived value. The various characteristics or evaluation parameters are entered in a chart as shown in Figure 5.13.

Assigning Parameter Weights

Once the maintenance manager has identified four or five characteristics by which the alternatives can be evaluated, a relative weight is assigned to each characteristic. A scale of one to ten is used to measure the importance of each factor. A high weight indicates great importance to the decision maker, a low score indicates less importance. For example, price may be weighted as an eight, maintenance cost as a six, and appearance as a four. These weight factors are entered in the column shown in the chart in Figure 5.13.

Situation: A gymnasium shower room has 100 shower heads which operate at a 5 gallon per minute flow rate.

Proposal: Install flow-restricting shower heads which operate at 2.5 gpm. Cost of one head including installation is $17.00.

Data: 320 showers are taken daily, 180 days per year
Average shower lasts 3 minutes
Heating shower water costs $.10 per 100 gallons
Water costs .90 per 1000 gals.

Solution: Compute the simple pay-back for installing 100 new flow-restricting shower heads.

Cost of Investment:
   100 Heads × $17.00 = $1700.00

Savings:
   320 showers × 3 minutes × 2.5 gallons saved
     = 2400 gallons per day saved
   Daily heating savings
     $.10/100 gal. × 2400 = $2.40
   Daily water bill savings
     $.90/1000 = $2.16
Annual Savings = $4.56 × 180 = $820.80
Simple Pay Back = $\dfrac{\$1700.00}{870.80}$ = 2.07 years

Figure 5.12

The weights of each alternative can now be compared. Each characteristic of each alternative has been given a score from one to ten, based on how well that alternative satisfies the specific characteristic. Next, the validity of the scores should be checked against each other. For example, if one score was twice the value of another, the question would be: "Does this alternative really satisfy this characteristic twice as well?" This process produces a more valid decision. The final step in the evaluation is to multiply each score by the weight of its respective characteristic. These products are then totalled for each alternative. The alternative with the highest score is the choice.

Sensitivity Analysis

When comparing several alternatives, each with several cost items, the reason why one is more economical than another might not be immediately obvious. Additionally, the quality of the cost estimates used in the evaluation may not be equal. The purchase price of a new piece of equipment, for example, may be exactly defined, while the expected failure rate and maintenance cost may only be loosely estimated. Yet the final decision may have been primarily the result of a small difference in maintenance costs between alternatives. This is determined by a *sensitivity analysis*.

Weighted Average Comparison Method

Problem: Determine the appropriate exterior wall system for a new building.

		Alternatives							
		#1 Brick		#2 Stucco		#3 Steel Siding		#4 Wood Siding	
Evaluation Parameters	Parameter Weight	score	wtd score	score	wtd score	score	wtd score	score	wtd score
Appearance	8	10	80	6	48	4	32	9	72
Initial Cost	6	4	24	6	36	8	48	7	42
Maintenance Cost	5	10	50	7	35	4	20	4	20
Durability	3	10	30	8	24	6	18	7	21
Permeability	7	7	49	8	56	9	63	6	42
Totals			233		199		181		197

Figure 5.13

To perform a sensitivity analysis, individual cost elements of each alternative are varied slightly or even to the highest or lowest extreme, and the effect of this change upon the decision is measured. For example, in cost comparisons using the present value method, the final decision may be very dependent upon the selected discount rate. In sensitivity analysis, the analysis can be performed again with a slightly higher or lower rate, and that effect measured. This determines the degree to which the final decision is governed by changes in any of the variables in the alternatives. If it becomes obvious that the final decision is based on a variable which is only roughly estimated, a more definitive estimate of that cost element should be prepared and the final decision re-examined and altered if necessary.

Summary

All maintenance decisions should be based on the cost of the alternative. All of the various elements of cost for any alternative must be accurately estimated and considered. Since these costs occur at varied times throughout the life of a facility or piece of equipment, the time value of money must also be considered. Each maintenance activity and decision represents an investment, ultimately in dollars. Therefore, maintenance costs can only be reduced through analysis of the dollar value of each maintenance project.

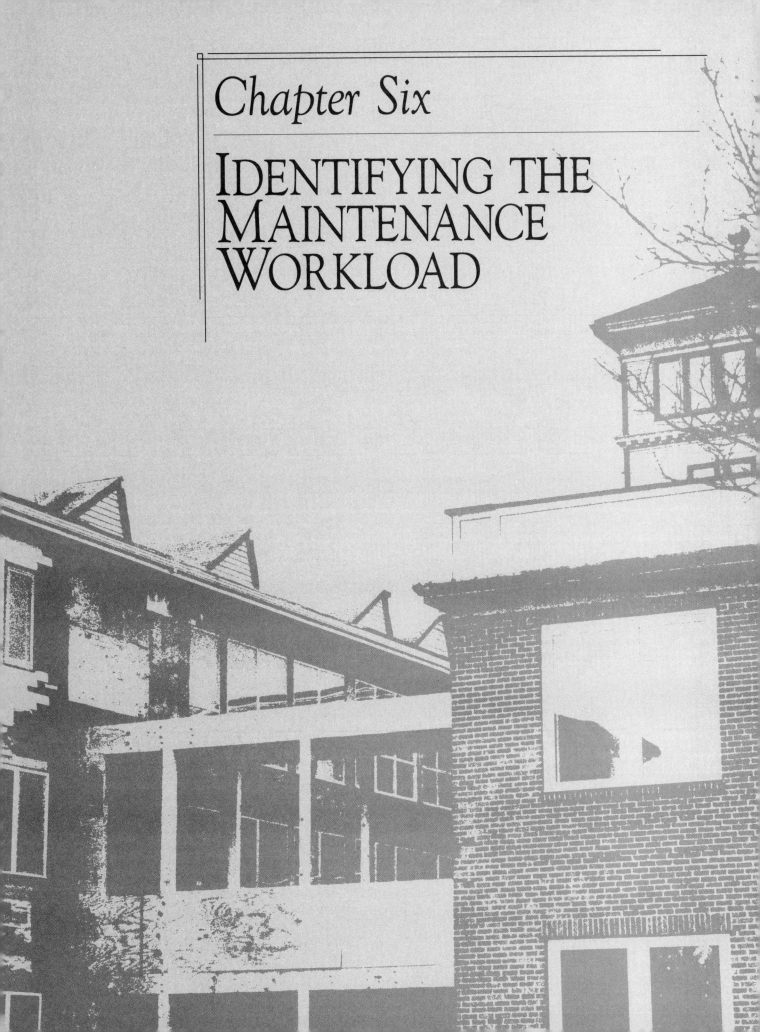

Chapter Six

IDENTIFYING THE MAINTENANCE WORKLOAD

Chapter Six

IDENTIFYING THE MAINTENANCE WORKLOAD

The identification of maintenance work does not, on the surface, appear to be a difficult task. Input can come from several sources including both the facility users and the maintenance staff. Simple user needs generate requests for repairs. Normal degradation of materials and equipment leads to replacement. Routine cleaning is necessary. The grass grows and must be mowed. If the maintenance staff were to simply attend to those obvious maintenance tasks, a sufficient workload would appear to keep that staff busy. However, the timely identification of the tasks necessary to allow continuous facility operation is an important factor in running a cost-effective, efficient maintenance program. The cycle of a typical maintenance activity is shown in Figure 6.1.

Maintenance work falls into two basic categories: *planned* and *unplanned work*. Planned work is that block of work which can be identified or predicted well in advance. Unplanned work covers those maintenance activities which appear unexpectedly and require immediate attention. Unplanned work falls into two general categories: catastrophic failures which arrest normal facility operations and small, labor-intensive repair jobs. A formal program to identify the full extent of both types of maintenance work should be established. Otherwise, failure to properly define the planned work may lead to an excess amount of unplanned work. This, in turn, disrupts the smooth execution of a maintenance program. Not all unplanned work can be eliminated, but any reduction of unplanned interruptions will reap desirable benefits. Planning work in advance allows the maintenance manager the ability to control maintenance rather than having maintenance control him.

Planned Maintenance Work

As previously stated, planned maintenance work includes those maintenance activities which can be identified and defined well in advance of execution. This advance identification allows the maintenance work force to be scheduled to fit the project needs and the availability of the maintenance work force.

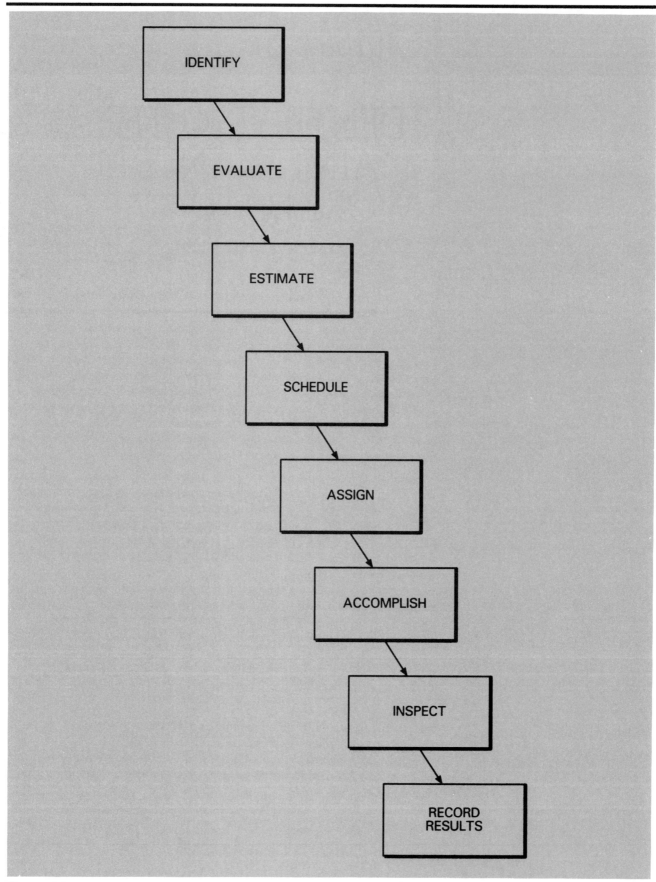

Figure 6.1

134

Recurring Work

Various maintenance tasks must be performed at regular and predictable intervals. For an existing facility, these recurring activities are identifiable using historical maintenance data, or past records. If a maintenance program and budget are being developed for a new facility, the recurring work should be determined by examining each area of the facility. Most recurring activities are related to either cleaning, groundskeeping, or the operation of the various facility systems. Many of these activities are mandatory tasks performed at a fixed frequency. Some tasks are mandatory, but the frequency is variable as dictated by management priorities, facility operations, or external influences. Still other tasks are completely discretionary, and may be delayed indefinitely.

The maintenance manager first determines which activities are *discretionary* and which are *mandatory*. The frequency of these activities is then categorized as *fixed* or *variable* and the most likely frequency of performance established. In an extremely austere budget, only the mandatory projects are performed and even those are performed only at the minimum frequency allowable. The results of this categorization are used by management to schedule the recurring work manpower and appropriate cost budgets.

Seasonal Work: Seasonal work is part of the recurring work performed on a building or facility. The performance of these activities is governed by season. The most obvious tasks in this category are landscaping related, such as lawn mowing, trimming, weeding, edging, and leaf pickup. The climate also effects the interior of a facility, primarily through the operation of heating or air conditioning systems. Most seasonal work dealing with these systems, however, falls under *preventive maintenance* (described in more detail later in this chapter and in Chapter 11). Figure 6.2 is a checklist to be used when analyzing a facility to determine the full extent of recurring and seasonal maintenance activities. The checklist is a generic listing for use in determining the basic activities common to most facilities.

For specialized facilities, the maintenance manager should examine specialized rooms, structures, and equipment not covered in Figure 6.2 to develop a supplementary list of maintenance activities for such spaces and equipment. Once generic and specialized lists of maintenance activities are identified, the estimated manpower and dollar requirements should be developed. (These estimates are prepared in accordance with the procedures and format shown in Chapter 4.)

Facility Scheduled Operations

There are many planned activities related to the operation of a facility. The facility's daily, monthly, or annual operation dictates both the need for and timing of various maintenance activities. For example, a college has a fixed date for major events, such as the opening of the semester or summer sessions, athletic events, and graduations. Each of these events requires some special preparation and work by the maintenance staff. For more frequently performed tasks, the schedule of facility operations may preclude the normal performance of maintenance activities during certain hours. In such cases, routine maintenance is relegated to off hours when the office workers have gone home, stores are closed, operating rooms are not in full use, or the classes are not in session.

Listing of Common Recurring Maintenance Work Activities

GROUNDS MAINTENANCE

Lawns
  Lawn and field mowing
  Trimming and edging
  Fertilizer application
  Insect and rodent control
  Leaf raking and removal

Planters, Trees, Shrubs
  Weed removal or control
  Annual plantings
  Trimming, pruning
  Insect and rodent control
  Fertilizer application
  General flower bed maintenance

Roads, Parking Lots, Walkways
  Street sweeping, cleaning
  Painting of traffic striping
  Sign maintenance
  Curb maintenance, painting

Miscellaneous, Site Work
  Litter collection
  Trash collection and disposal
  Painting fences, small buildings
  Exterior lighting (bulb replacement)
  Cleaning out storm drains

BUILDING EXTERIORS

Walls
  Painting
  Cleaning, mildew, mold, fungus removal
  Caulking renewal

Windows
  Window washing
  Storm window/screen installation/removals

Doors
  Cleaning, washing
  Storm/screen door changeouts
  Lubricate hinges, locks

Roof
  Gutter and downspout cleaning
  Roof drain cleaning
  Trash removal off flat roofs

Floors
  Vacuuming and sweeping
  Washing and waxing
  Stripping wax
  Mopping
  Carpet steam cleaning

Figure 6.2

Special work requirements do arise which disrupt the normal flow of maintenance work. These place a one-time burden on the maintenance work force. However, by identifying the special events and regular restrictions associated with facility operation in advance, the maintenance manager has the opportunity to schedule the work force to meet these requirements.

Preventive Maintenance

Preventive maintenance activities account for a major portion of the planned maintenance workload. The determination of the need for and frequency of preventive maintenance activities is a critical process. Preventive maintenance work activities are analyzed and identified by both the maintenance supervisor and the maintenance work force. Care must be taken, however, as too little preventive maintenance results in equipment failure, and too much wastes money. (See Chapter 11 for an analysis and development of a preventive maintenance program.) The activities which result from that analysis, coupled with the recurring, seasonal, and operation-related maintenance work, form that portion of the entire maintenance workload for which firm planning and scheduling can be performed.

Listing of Common Recurring Maintenance Work Activities (continued)

Walls, Ceilings
Dusting
Cleaning, spot cleaning
Touch-up painting

Doors, windows
Cleaning, washing
Drape, shade, curtain cleaning
Touch up painting

Public Spaces
Replace paper towels
Replace toilet paper
Replace soap — cake, liquid, powder
Empty trash receptacles
Empty/clean public ash trays

Offices, Shops, Workplace
Waste collection
General dusting
Vacuuming
Sweeping
Wash & wax floors

Figure 6.2 (continued)

Unplanned Maintenance Work

If all maintenance work was identifiable in advance, all maintenance workers would be efficiently scheduled, all maintenance costs accurately predictable, and maintenance management less of a challenge. But things break, parts wear out, and facility users change their minds, all with some frequency. And every once in a while, catastrophies occur. Any of these events sends a disruptive pulse through the maintenance organization. Although it is easy to assume that some such events will occur at some probable frequency, the exact character of these unplanned maintenance requirements is not completely predictable. It is necessary, therefore, to understand the means by which this unplanned portion of the maintenance workload evolves so that a system may be devised to deal effectively with this workload.

Unplanned work varies daily and is reported by numerous sources, each event varying in scope and urgency. Four parties input unplanned maintenance work into the system: the facility user, the maintenance work force, the maintenance manager, and the facility and its systems. Each of these roles is discussed in the following sections.

User-Noted Discrepancies

Facility users are particularly adept at identifying immediate changes in the condition of the facility in which they work. Some facilities managers feel the user is a hindrance to the running of a "smooth" maintenance program. However, the input from the facility user is actually the most important and valuable input available to the maintenance manager. The facility users represent hundreds of eyes and ears that can alert the maintenance staff to deficiencies that encumber the ability of the facility to perform its intended function.

The facility user generally identifies only real and continuing problems. Facility users rarely have the expertise to recognize an impending failure. For example, if the light in their office is slow to energize, but eventually burns brightly, the user most likely ignores that minor inconvenience until it fails, at which time they will react quickly and seek assistance. Figure 6.3 outlines the types of facility discrepancies likely to be reported by facility users.

Facility user and maintenance staff observations are not discretely different. However, the facility manager should have a clear understanding of which discrepancies the user will identify in order to plan the maintenance workload. Without input from the user, the maintenance work force would spend inordinate amounts of time on daily inspection visits.

Maintenance Work Force Input

The maintenance workers are intimately familiar with the facility, its systems, and its equipment. Because they are frequently working in one area or with some particular system, maintenance workers can readily note changes in the performance of the facility. The maintenance worker has the expertise, in many cases, to note the degradation in performance of a particular portion of a facility prior to its actual failure.

As maintenance workers walk through various parts of the facility, they are stopped by the facility users and advised of a problem. In general, the facility user relies on this informal contact as being sufficient to have notified the maintenance department of the problem. Although disruptive to the worker's intended mission, this form of problem identification has a distinct advantage. If the problem is within the worker's area of expertise, the worker may be able to rectify the problem immediately. Even if the time required exceeds the time available to the worker, the worker can at least perform a preliminary investigation. This information, over and above that which a facility user might provide, results in a clear statement of the problem and probable action required to remedy the discrepancy. This clarification allows the planner/estimator to prepare a reasonably accurate estimate and enables the shop foreman to budget the proper amount of time for the repair. The disadvantage of this type of work identification is that the worker might forget to notify the maintenance control staff of the problem. The facility user believes that the solution or repair is forthcoming, but the maintenance management staff is not even aware of the problem.

In addition to being advised of problems by the facility users, the maintenance worker personally notes problems while walking through various parts of the facility. In such cases, the report of the problem to maintenance control includes the informed input of the worker described above. The maintenance worker also becomes aware of facility or equipment problems during the direct performance of a maintenance project. When a piece of equipment is partially disassembled for maintenance, the worker may notice parts which are excessively worn, out of balance, or improperly adjusted.

As with the casually noted problems, many of these discovered discrepancies can be repaired immediately. Additionally, if the maintenance manager or shop foreman has reason to suspect that problems might be discovered, the maintenance worker can be

---

User Reported Problems

Specific Work Environment
Burned out light bulbs
Lack of heat/air conditioning
Lack of ventilation
Malfunctioning electric circuits
Lack of janitorial services
Any safety discrepancy
Unusual noise, odors

General Facility Environment (Interior)
Malfunctioning elevators
Clogged toilets, missing paper products
Malfunctioning locks, doors, windows
General cleanliness

General Facility Environment (Exterior)
Landscaping appearances
Roads and walks — cracks, potholes, etc.
Icy, snow covered walks

Maintenance Support Spaces
Unusual noises, odors

Figure 6.3

forewarned. The preventive maintenance order (outlined in Chapter 11) notes particular areas to check, contains instruction for the immediate repairs, and describes the method for reporting the need for later repairs which may be beyond the worker's immediate capability to rectify.

The quality of information which accompanies the report of a worker-discovered discrepancy usually enhances the ability to estimate and schedule the repair. This is not, however, always the case. Since maintenance workers are production-oriented, the reports they generate are apt to be expressed as statements of the required *solution* rather than a description of the *underlying problem*. The shop foreman and maintenance manager must, therefore, consider very carefully the particular worker's ability to properly diagnose problems of the type for which he or she has proposed a solution. In some cases, it may be necessary to question the worker to determine the underlying problem which caused the worker to recommend the particular solution. Figure 6.4 illustrates the path of actions which lead to the final addition of a maintenance worker discovered project to the maintenance work backlog.

Improvement Recommendations: The maintenance worker provides valuable work input through the recommendations of new or revised maintenance procedures. Sample recommendations likely to be developed and submitted by maintenance workers are listed below.

- Purchase of new diagnostic tools.
- Revised procedures for preventive maintenance.
- Expanded investigation of problems by other craftsmen.
- Replacement of difficult to maintain equipment.
- Purchase of equipment to save time in repairs.
- Rectification of a safety problem area.
- Revise methods to those of a worker's previous employer.

All of these types of suggestions should be evaluated by the maintenance managers as potential improvement projects. When a worker proposes a new or revised procedure that involves the expenditure of funds, either to purchase equipment or expand labor requirements, the proposed benefit must be identified and quantified in the manner shown in Chapter 5. If a decision is made to move forward with a worker suggestion, management must follow through to ensure that the proposed benefits are realized.

Inspections

An inspection is a purposeful examination of a facility with the intent to identify problems. The key to the success of inspections lies in the dedication of specific time to that specific purpose. No intention is made during an inspection to immediately repair the identified problems; the inspector merely identifies them. Inspections are much more likely to uncover long-term trends in facility degradation than the continuous observations of the user or maintenance worker. The facility user and maintenance worker can both become comfortable with a facility and fail to note minor incremental changes in facility conditions. The inspection, conducted formally and less frequently than daily observation, brings long-term changes into view and is an essential component of the entire set of procedures for identifying the full extent of necessary maintenance.

Baseline Facility Surveys: The most time consuming, yet most productive, type of inspection is the *baseline facility survey*. This is a thorough examination of the entire facility. Its purpose is to determine the exact condition of the facility at the time of the inspection. The results of that inspection become both a statement of needed maintenance work and form a baseline against which the next scheduled inspection is measured. Two key factors influence a baseline inspection: what is inspected and who performs the inspection.

Since the baseline inspection is an intensive look at a broad variety of components, a *checklist* is often used to ensure full coverage during the inspection. The checklist is a reminder of areas and items to be inspected. While the baseline inspection should cover a broad area, it must also be somewhat superficial; extensive amounts of time should not be spent in the exact definition of the scope of each noted problem area. A list of work items is the result of the inspection, including both specific projects and those which need more intensive investigation. The focus of the inspection should include:

- Areas not normally visited by users or workers.
- Items which degrade slowly and progressively and are unlikely to be noted casually.
- Items requiring more expertise than available to the layman.
- Specific items critical to the safety of the facility users.
- Closer looks at "weak links" which could lead to costly or catastrophic events.

Problems Reported by Maintenance Workers

Discrepancies in equipment performance noted during routine preventive maintenance
 Required checks
 Other observations on serviced equipment
 Observations on other equipment in same space

Casually noted discrepancies
 Most of user noted items but concentrated within the workers trade specialty
 Items reported by users directly to worker in lieu of trouble desk

Improvements to maintenance methods
 Worker suggested projects which will improve equipment performance or reduce maintenance

Figure 6.4

The success of inspections is somewhat dependent on the structure of the checklists. Inspection checklists should provide a focus for each area. Since the inspection covers all physical areas and component systems of a facility, it is easy to forget to look at each specific area within each different room or building. Figure 6.5 is an example of a *master inspection checklist* for an entire facility. It is limited to those areas which will not be specifically covered by the sample *individual building checklist* shown in Figure 6.6. The individual building checklist includes items peculiar to the entire building and not specifically covered by the *individual space checklist*, an example of which is shown in Figure 6.7. Figures 6.8 and 6.9 are individual checklists for mechanical and electrical systems, respectively.

Completion of the baseline inspection is a time-consuming procedure. Individual facility maintenance managers must determine the exact frequency of these full inspections; a period between inspections of not greater than every two years is suggested. For facilities which are subject to intensive use, more frequent inspections may be necessary. The structure of the various checklists allows for a staggered inspection of an entire facility. The ongoing pursuit of a maintenance program does not usually allow full-time dedication to such an inspection from start to finish. However, individual checklists can be filled out independently. A complete inspection will include: a completed checklist for the entire facility (see Figure 6.5); one completed checklist for each major building (see Figure 6.6); one checklist for mechanical systems per building (see Figure 6.8); one electrical system checklist per building (see Figure 6.9); and one space checklist for each major space within a building (see Figure 6.7). For the sake of time and simplicity, the individual spaces, for which a single checklist is completed, can be combined into logical groupings by function of the space and one checklist completed for that grouping. For example, an office building might have only three individual space checklists. One might address public spaces such as lobbies, corridors, and rest rooms. The second might address all office spaces, while the third might cover special purpose rooms such as computer rooms.

As the inspector completes the various checklists, maintenance work items are identified. Each noted discrepancy must then be recorded and added to the maintenance backlog. The inspector may take only brief notes during the inspection and fill out more detailed maintenance work requests later. Or, the inspector may find it easier to simply fill out formal work request forms as items are noted. This is a matter of personal preference. Delaying the writing of the work request allows for a quicker inspection, but adds the risk that the inspector may not be able to fully recall all data needed for the work request when back in the office.

In addition to providing a dedicated focus for the inspector, the scheduling of a formal inspection also provides stimulus to the maintenance work force and facility users to bring the facility up to its best condition. An unannounced inspection generally yields an accurate picture of the facility's day-to-day condition, while a scheduled inspection spurs the maintenance work force to bring the entire facility up to its best possible condition. Each of these options has validity as a maintenance management tool and the choice is left to the maintenance supervisor.

**OVERALL FACILITY
INSPECTION CHECKLIST**   DATE _____

Grounds

  Landscaping

    Lawns

      Properly maintained _____

      Fertilized _____

      Bug infested _____

      Properly sloped, drained _____

      Properly mowed, edged, trimmed _____

    Planting beds

      Weed free _____

      Trimmed _____

      Growing _____

      Fertilized _____

      Edged _____

    Shrubbery

      Trimmed _____

      Growing _____

      Infested _____

    Trees

      Need pruning _____

      Any rotting _____

      Infestation _____

Exterior Construction

  Fencing

    Intact _____

    Painting needed _____

    Rusting _____

    Gates operable _____

    Gates lockable _____

  Signs

    Plumb, level _____

    Visible _____

    Faded _____

    Properly lighted _____

  Minor Buildings, Sheds, etc.

    Painted _____

    Trimmed _____

    Doors operable, lockable _____

    Park benches, etc. intact, safe _____

  Utility enclosures

    Transformers enclosed _____

    Properly marked, hi voltage, etc. _____

    Sewage manholes intact _____

Figure 6.5

Paved Surfaces

  Roads

    Surface intact _____

    Sloped for proper drainage _____

    Storm drains clear _____

    Markings, striping clear _____

    Potholes, cracking _____

    Joint sealants _____

    Curbs intact _____

    Crosswalks provided _____ Marked _____

    Traffic control signs present _____

  Parking Lots

    Properly marked _____

    Curbing in intact _____

    Handicapped spaces provided _____

    Drainage proper _____

    Guardrails needed _____ Intact _____

    Walkways provided _____

  Walkways

    Safe, tripping hazards _____

    Drain properly _____

    Handicap access provided _____

    Joints intact _____ Sealed _____

Exterior Lighting

  Street Lighting

    Lights in proper location _____

    Light poles in sound condition _____

    Lighting fixtures intact _____

    All bulbs burning _____

  Walkway Lighting

    Lights functioning _____

    Dangerous shadow areas _____

  Building, Sign, Misc. Lighting

    Functioning _____

    Properly aimed, aligned _____

    Entrance lighting intact _____

    Security lighting provided _____

    Glare problems caused _____

Comments: _____

_____

_____

_____

_____

_____

_____

Figure 6.5 (*continued*)

**GENERAL BUILDING
INSPECTION CHECKLIST**   DATE _____

Building Exterior

Walls

Sound condition _____

Properly surfaced _____ Painted _____

Discolored _____ Stained _____

Caulking condition _____

Mold, mildew, fungus present _____

Vines, ivy, etc. present _____ Damaging _____

Any visible cracking _____

Any perceptible deformation, deflection _____

Windows

Glazing in good condition _____

Caulking sound _____

Frame condition _____ Rust _____ Peeling paint _____

Glazing distorted _____

Screens attached _____ Ripped _____

Window Walls, Storefronts

Glazing in good condition _____

Caulking sound _____

Entrance Doors

Hardware operational _____

Kickplates present _____

Glazing intact _____ Distorted _____

Panic hardware operational _____

Doormats provided _____

Properly marked _____

Exit signs present _____ Operational _____

Roof

General appearance satisfactory _____

Drainage piping intact _____

Gutters, downspouts _____

Flashing intact _____

Soffit vents clear _____

Ventilators, waste vents flashed _____

Attachments (weathervanes, etc.) secure _____

Flat roofs

Alligatoring _____

Bubbles _____

Ponding water _____

Drains clear _____

Scuppers, overflows available _____

Perimeter flashing intact _____

Penetrations flashed _____

Pitch pockets full _____

Gravel smooth _____

Figure 6.6

Exterior stairs, ramps

Tripping hazards _____

Properly lighted _____

Handrails secure _____ Painted _____

Building Interior

Public spaces

Corridors, lobbies

Floor condition _____

Tripping hazards _____

Proper flooring materials _____

Walls clean _____ Painted _____

Directional signs provided _____

Ceilings clean, intact _____

Exit signs visible, operational _____

Rest Rooms

Clean _____

Mildew, mold, fungus present _____

Ventilation proper _____ Odors _____

Fixtures intact, clean _____

Toilet partitions intact _____

Lighting proper _____

Handicapped facilities provided _____

Leaks, drips, running toilets _____

Stairs, stairwells

Free from tripping hazards _____

Exits properly marked _____

Lighting proper _____

Handrails secure _____

General cleanliness satisfactory _____

Mechanical & Electrical Equipment Spaces

Cleanliness satisfactory _____

Doors locked, lockable _____

Safety notices posted _____

Ventilation satisfactory _____

Drips, leaks, etc. _____

Excessive odors, smoke, haze _____

Improper materials stored in space _____

All wiring secured _____ In conduit _____

Any flammable materials stored _____

General review for fire hazards _____

Unoccupied, storage spaces

Cleanliness satisfactory _____

Flammable materials stored _____

Any apparent safety hazards _____

Any apparent fire hazards _____

Materials or supplies stored safely _____

Fire Alarm System

Panel properly visible _____ Operational _____

Figure 6.6 *(continued)*

**INDIVIDUAL ROOM
INSPECTION CHECKLIST**

DATE _____

Complete one for each room or group of similar rooms

Floors

Material in proper condition _____

Well maintained _____

Stain, marks, scratches, dents, etc. _____

Tripping hazards _____

Visible sagging, settlement _____

Walls

Wall covering, finish proper for room _____

Wall covering, paint etc., in good condition _____

Watermarks, condensation on exterior walls _____

Trim, chair rails, moldings need painting _____

Dents, scratches, etc. _____

Doors

Proper hardware installed _____

Hardware in good condition _____

Door surfaces in good condition _____

Kickplates installed _____ Needed _____

Exits properly marked _____

Entrance signs, directional signs in good condition _____

Windows

Glass sound, clear, free from distortion _____

Frames need painting, repair _____

Operable windows work properly _____ Freely _____

Shades, drapes, blinds, etc. installed properly _____

Electrical Services

Sufficient outlets _____ Proper voltage _____

All wiring secured _____ In conduit _____

Outlets and boxes covered _____

Telephone Systems, etc.

Sufficient outlets _____

Wiring secure _____ Not tripping hazards _____

Environmental Matters

Lighting

Fully operational _____

Adequate for room use _____ Excessive _____

Proper switches installed _____

Heating/Ventilating/Air Conditioning

Temperature proper _____

Any cold/hot spots _____

Thermostat installed _____ Proper setting _____

Draftiness _____

Insufficient ventilation _____

Figure 6.7

147

**INSPECTION CHECKLIST
FOR MECHANICAL SYSTEMS**

DATE _____

Heating/Cooling Systems

    Does system provide adequate heat/cooling _____

    Is heated/cooled air evenly distributed _____

    For Furnace Systems

        Do furnace emissions meet EPA standards _____

        Any notable fuel leaks _____

        Notable leaks from exhaust stack _____

        Records show proper servicing _____

        Proper safety valves, pressure reliefs _____

Water System

    Water pressure adequate _____

    Water free from noticeable odors, tastes _____

    Is main cutoff valve operable _____

    Any noticeable leaks, excessive condensation _____

    Water heater set to proper temperature _____

    Water heater provided with pressure relief _____

Sanitary System

    Piping free from leaks _____

    Flow apparently sufficient _____

    Cleanouts accessible _____

    Inspect drain/absorption fields _____

Ventilation Systems

    Ventilation adequate in all spaces _____

    Fans noisy _____ Excessive vibration _____

    Intake filtering systems clean _____

    Exhaust outlets screened, clean _____

    Exterior louvers, etc., secured in place _____

Exterior Fuel Tanks

    Tank marked for contents _____

    Above ground tanks enclosed in dikes _____

    Buried tanks have leak monitoring systems _____

    Tanks properly grounded _____

    Noticeable leaks _____

    Need painting or other maintenance _____

Comments: _____

_____

_____

_____

_____

_____

_____

Note:

The above list is intended to allow a non-engineer to perform a basic inspection of mechanical systems. In order to properly ascertain the condition of all mechanical systems, the characteristics of each system (temperatures, pressures, emission contents, etc.) will have to be determined and compared with design standards. Additionally, all maintenance records must be examined to determine if proper levels of maintenance are occurring.

Figure 6.8

**INSPECTION CHECKLIST
FOR ELECTRICAL SYSTEMS**

DATE _____

Exterior Electrical Service

Are transformers owned by facility or utility _____

If self-owned, have they been tested annually _____

Any signs of leaks from transformers _____

Any signs of electrical arcing, burning _____

   Aerial wiring

   Consistent "droop" for all wire _____

   Poles, insulators in good condition _____

   Guy wires provided _____ Poles plumb _____

   Transformer enclosures secure, well marked _____

Interior Electrical Distribution System

Incoming conduit properly marked _____

Main switch panel identified _____

Circuit breakers in main panel marked _____

All panel boards, junction boxes covered _____

All wiring in conduit _____

Conduit properly secured to walls, etc. _____

Panels throughout buildings marked _____

Panel schedules posted inside panel doors _____

Any missing breakers, open spaces in panels _____

Wiring, Outlets, Wall Switches, Lighting Fixtures

All cover plates in place _____

All junction boxes covered _____

All wire in conduit _____

Fixtures operate properly _____

Sufficient outlets provided _____

Tripping hazards present _____

GFI circuit breakers in baths, kitchen, etc. _____

Grounded wiring (3 prong) used for portable electrical appliances _____

Emergency Circuits

Emergency generator used _____

Automatic start and switchover functional _____

Are proper circuits on generator _____

Fuel sufficient for predicted outage _____

Generator tested regularly _____ Under load _____

Emergency lighting provided _____ Battery powered _____

Comments _____

_____

_____

_____

_____

_____

Note:

Formal inspections of electrical systems must be performed by licensed electricians or electrical engineers. This listing is intended for use by the non-engineer. It is not a substitute for a thorough inspection.

Figure 6.9

The success of any inspection program also depends on the assignment of the appropriate inspector; this person should be somewhat removed from the responsibility for direct performance of maintenance work in order to have a reasonably impartial perspective. The electric shop foreman might be able to perform a thorough inspection of the facility's electrical system, but an electrical engineer can provide an outside perspective which may uncover some routine, but unacceptable, condition. As a rule, the inspector should hold proper expertise in the inspection area, but should be working at least one level higher in the organization than those directly supervising or performing that type of maintenance work. In small organizations, there may not be such an individual who is sufficiently removed from the performance of work to render an unbiased viewpoint. In those cases, outside consultants or engineers may be used.

Some inspections are performed by agencies outside of the facility organization. Fire safety inspections, OSHA site visits, EPA investigations, sanitation inspections, or other inspections necessary for licensing or accreditation requirements generally produce unplanned maintenance work items. In some cases, the work required by these types of inspections is of greater urgency if the facility operation is limited or encumbered by the unsatisfactory condition.

Ad Hoc Inspections: In addition to the formal scheduled inspection, the maintenance manager may perform unscheduled inspections of specific areas of the facility at frequent intervals. The need for these inspections may arise from reports that maintenance performance is lacking in some area or by some specific group of maintenance workers. For this reason, the *ad hoc inspection* is a quality control tool. It provides information concerning the performance of a particular type of maintenance or of the condition of a specific area of the facility. Formal checklists are not necessary for ad hoc inspections, since the inspector is usually aware of the areas that should be examined. The results of an ad hoc inspection include the identification of new maintenance work items and revised procedures for maintenance.

Management Staff Input

All of the previous methods of identifying unplanned maintenance work have been *reactive*. The physical condition of the facility has dictated the need for work, whether reported by a facility user or identified through an inspection. In the interest of providing as much advance notice of any maintenance problems, the maintenance manager should also attempt to predict when certain unplanned maintenance requirements will arise. This section contains methods for *predicting* unplanned maintenance requirements.

Trend Analysis: The maintenance manager can usually identify trends in the maintenance of specific areas without a detailed analysis. For example, the repeated occurrence of problems with air conditioning units might indicate the probability that more such failures are likely to occur. It might also indicate a deficiency of the air conditioning maintenance program. The shop foreman may report that the ratio of planned to unplanned work in a certain area has changed, finding the scheduling of workers constantly interrupted by emergency response. Once such a trend is identified, several possible actions result. The maintenance manager might schedule a complete inspection or servicing of all air conditioning units. The shop foreman might investigate the problems in detail to determine if some specific procedure performed (or neglected) by a worker was the cause for the increased failure rate.

Sympathetic Maintenance: When a specific failure occurs, the maintenance manager must decide whether the failure was inevitable or symptomatic of a basic problem with the particular unit or with the maintenance procedure. No matter what the cause of failure, the maintenance manager must anticipate that similar problems are likely to occur in similar units or systems installed elsewhere in the facility. Therefore, a single failure may result in the need to examine all like units for indications of similar problems. This immediate examination of other units is unplanned and performed only to preclude the more severe costs of a similar failure. This is known as *sympathetic maintenance*. The maintenance action indicated may be a one-time repair to several units or a new procedure added to the previously identified preventive maintenance activity for those units.

Improvements: A final unplanned maintenance work item is the *improvement project*. As changing technology produces new equipment, tools, materials, and maintenance methods which produce a positive pay-back, the maintenance manager may desire to implement that improvement immediately. This is particularly true for improvements with very short pay-back periods (see Chapter 5 for a more detailed discussion of pay-backs).

The manager must decide whether to implement the project immediately or to schedule it for a later date. This decision is governed by the availability of funds, the implied savings, and the desirability of disrupting the planned schedule of maintenance work to perform the project.

Modifications

Modification projects are unplanned work inputs. They are generally requested by the facility user and are often tied to a specific schedule of facility operations. Such projects must be performed within the constraints of the facility and not necessarily at the convenience of the maintenance schedule. Remembering that one objective of the maintenance manager is to provide stability to the maintenance work schedule, the time for accomplishing modification projects must be negotiated with the requesting facility user. If sufficient time is not available to perform the modification project or the project is quite large, the use of contracted services should be considered.

Figure 6.10 is a tabular summary of the planned and unplanned inputs to the maintenance work backlog.

The Work Identification Process

The maintenance manager must establish a formal process for accepting these inputs from the various methods by which maintenance work is identified. This process must be convenient, easily followed, and reliable. The primary reason that problems in a facility are not addressed is not that the problems go unnoticed but, rather, that the persons who note the problems simply do not know who to tell. If the facility user is unaware of how to access help from maintenance, minor problems will go unreported and unresolved, leading to more costly failures. The same problems result if the maintenance staff does not have a *specific procedure* to follow to report their discovered discrepancies. Two key elements are necessary to develop a work identification process. The person discovering the discrepancy must have a *specific person* to contact, and the report of the discrepancy must provide *sufficient information* for the maintenance management staff to evaluate and plan the proper maintenance action.

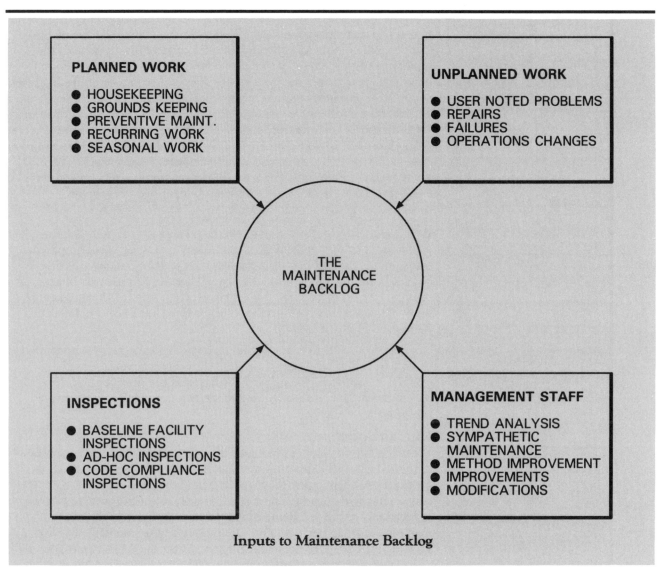

PLANNED WORK

- HOUSEKEEPING
- GROUNDS KEEPING
- PREVENTIVE MAINT.
- RECURRING WORK
- SEASONAL WORK

UNPLANNED WORK

- USER NOTED PROBLEMS
- REPAIRS
- FAILURES
- OPERATIONS CHANGES

THE
MAINTENANCE
BACKLOG

INSPECTIONS

- BASELINE FACILITY
  INSPECTIONS
- AD-HOC INSPECTIONS
- CODE COMPLIANCE
  INSPECTIONS

MANAGEMENT STAFF

- TREND ANALYSIS
- SYMPATHETIC
  MAINTENANCE
- METHOD IMPROVEMENT
- IMPROVEMENTS
- MODIFICATIONS

Inputs to Maintenance Backlog

Figure 6.10

Work Reception

Every maintenance organization, regardless of its size, should have a formal method for facility users to request maintenance assistance. In an apartment building, the tenants usually contact the "maintenance man" for help. In a motel or hotel, the users are guests who call the front desk for assistance. In larger facilities, the users might know the name of the primary maintenance worker and will call that person with problems. For a large facility, the person designated to receive these requests is known as the *work receptionist* or *trouble desk*. The phone number for this individual must be readily available to all personnel. The work receptionist is a critical link in the overall maintenance organization, as this position is the funnel through which all requests should pass. The work receptionist should possess the following capabilities and personal traits:

- an understanding of the facility layout,
- an understanding of the facility's operational staffing,
- a basic understanding of typical maintenance problems,
- patience to deal with frustrated facility users,
- dedication to immediately record problems,
- sufficient judgment to discern routine from emergency repairs,
- a helpful pleasant nature,
- an assuring, professional phone presence, and
- genuine sympathy to the requesting party's problems.

This list should be considered when evaluating applicants for the work receptionist position.

The facility user who reports a problem must feel that:

- the problem has been acknowledged,
- the work receptionist understands the scope of the problem,
- that the problem was worth reporting, and
- that help is likely to be forthcoming.

The maintenance manager should ensure that the work receptionist regularly leaves callers with this impression.

When requests yield no response, the frequency of such requests falls off. Therefore, the second key function of the receptionist is to convey the stated problem to the maintenance control branch with sufficient information to allow for proper maintenance response. This is done in the form of a formal *work request*.

The Work Request

The work request is the vehicle by which work is formally identified and conveyed to the maintenance worker who eventually performs the desired repairs. Work requests are received in both oral and written form. If written, a specific published form should be used. If the work request is submitted to the work receptionist, the same form should be fully completed by the work receptionist and passed along to the appropriate party. The work request form should provide proper space for the reporting party to provide the following information:

- **What:** What is the exact problem. There is a tendency to request specific actions or solutions rather than stating the underlying problem. The work receptionist should be persistent in determining and defining the true problem in addition to accepting suggested solutions.
- **Where:** The location of the problem must be defined. Depending on the size of the facility, this may include the building name, floor, room number, or other necessary information to enable the maintenance worker to locate the problem.

- **Who:**   The name of the person making the request or the person to contact (if different). Also, a phone number where they may be reached. This is used by the maintenance staff if further information is required, by the maintenance worker when making the repairs, and for obtaining a signature approving the work when completed. In addition to the name of the individual, the title of the organizational element to which the requesting party belongs should be noted. If the requested work is conducted on a reimbursable basis, the approving official and proper accounting data should be given to authorize and apply the charges for the work.
- **When:**   The date that the request is made should be entered and the time desired for completion. Also, any peculiar scheduling limitations which should guide the time of execution of the project should be spelled out.

Figure 6.11 is a suggested work request form for use in requesting maintenance services. The form is utilized by the facility user in requesting the work, by the maintenance staff in preparing estimates and providing guidance for scheduling and procedures, and by the maintenance worker to record the results of the repairs.

The work request is normally filled out first by the initiator of the request. The work receptionist fills out work requests for those requests made over the phone. If the maintenance records are to be kept manually, several copies of the work request form are needed to track the project through to completion.

The Maintenance Work Backlog

The *maintenance work backlog* consists of all work identified as necessary but awaiting completion. Much of this work load is planned work and the staffing necessary to complete that planned portion can be developed with a fair degree of accuracy. The remaining portion of the backlog consists of unplanned work. If the staffing is of sufficient size to accommodate the accomplishment of every request immediately, there is no backlog of unplanned work. This situation rarely happens, nor is it always desirable. If the staff is of such a size that they were standing by waiting for the next service call, maintenance costs would be unacceptably high. The backlog is important, therefore, to provide a pool of work sufficient to keep the maintenance workers fully employed. The backlog should be divided into categories of urgency and also into separate listings for each shop or maintenance element. The management of the backlog and its size is covered in greater detail in Chapters 7 and 8.

Summary   Maintenance work is identified by several sources. The maintenance manager should ensure that reporting procedures are available for each of these various parties to relay the need for work to the maintenance staff. This is normally accommodated through a single point of contact on the maintenance staff, known as the work receptionist. The maintenance manager should also ensure that long-term changes in the facility are noted through the use of regularly scheduled inspections. The accumulated listing of all work awaiting completion is known as the maintenance backlog.

REQUEST FOR MAINTENANCE SERVICES

Maintenance Control Number|_____
Description of Requested Services

Desired Completion Date_____

Work Location		Requested By	Phone	Date Submitted
Bldg.	Room			

For Reimbursable Work Only	Authorizing Official
Account to be Charged	

BELOW PORTIONS TO BE COMPLETED BY MAINTENANCE CONTROL BRANCH

Priority	Estimated Man-hours	Estimated Mat'l.Cost	Approved By
			Date

Special Instructions

Completion Data	Remarks
Man-hours Expended Mat'ls. Used	

Work Completed By	Date Completed	Customer Approval

Figure 6.11

Chapter Seven

Evaluating and Executing Maintenance Work

Chapter Seven

EVALUATING AND EXECUTING MAINTENANCE WORK

Each individual work request, whether initiated by the facility user, maintenance worker, or maintenance management staff, must be evaluated and assigned a priority for accomplishment. (See Chapter 6 for a complete description of the work request.) The criteria for evaluation are drawn from such questions as: Is the work necessary to maintain operations? Will the work enhance profitability? Will the work delay or prevent more costly work? Will the work reduce long-term operating costs? Is the work necessary to abate a safety problem? Can the work be delayed without risking more costly repairs? The maintenance manager first determines the urgency with which the maintenance staff must respond to the work request. Once the urgency is determined, the work request is scheduled for accomplishment, assigned to a craftsman, and executed. The evaluation, scheduling, assignment, and execution of the work are each actions guided by management policies and decisions, the methodology of which should be clearly defined.

Evaluating Maintenance Work

The evaluation of any potential maintenance work activity involves several considerations: the propriety, need, value, and urgency of each proposed activity. These factors are each considered and satisfied when planned work is initially identified. Therefore, planned work (preventive maintenance, recurring work, seasonal work, and that work necessary to support specific facility operations) does not require a second evaluation before scheduling.

Unplanned work, requested in the form of a work request, is not automatically scheduled. Since the input comes from outside of the maintenance management staff, it must first be carefully evaluated against the previously mentioned criteria.

Each organization has numerous standards which determine what types of maintenance work will be performed. Internally, the organization has specific policies concerning safety procedures and authorized types of construction. There are also various sets of laws, rules, codes, and regulations, external to the organization, which allow or prohibit various types of construction or equipment within any facility. These include requirements of the Occupational Safety and Health Act (OSHA), regulations published by the Environmental Protection Agency (EPA), regulations published by the state equivalents of both OSHA and EPA, local building codes, local zoning regulations, and fire safety and protection codes.

Propriety

Unless the requested repairs would significantly change a piece of equipment from its pre-existing functions, the question of propriety does not usually apply. When considering the propriety of any requested modification or improvement work, however, the maintenance manager considers the question, "Is this work allowed?" If the work is allowed, the question of whether or not the requesting party is authorized to request such work must be answered. In modification projects, for example, there will be reimbursement for the work. In such cases, the work should not start if the requesting party has no authority to commit the funds for such work.

A final factor in evaluating the propriety of the work is to determine whether the proposed solution is the proper course of action. The requesting party may have worded the work request in the form of a problem solution, rather than a stated problem. The maintenance manager must determine the underlying problem which has prompted the request and evaluate the various options available. This may require engineering input. A request to install a window air conditioner to augment a permanently installed system which is not providing proper cooling pinpoints a legitimate problem, but perhaps suggests a poor solution. An engineering analysis may indicate that the permanently installed system could be balanced or adjusted to provide the needed cooling at far less expense than the installation of a window unit.

Need

In some cases, a request is made for no apparent reason. The initiator may not have fully identified the underlying cause that generated the request. For example, a facility user who has a problem with an electrical appliance, might request that the electrical circuits be checked. In this case, a phone call to the user may reveal that the problem has already been isolated to the circuit and that the requested work should proceed. On the other hand, the phone call may reveal that the user has not eliminated the appliance as the cause of the problem and, therefore, there is no reason to suspect the building electrical circuits.

More often, the question of need becomes muddled with the question of urgency. Some work may be necessary, but not of any particular urgency. In such cases, the facilities maintenance manager should consider the following questions: Is the work necessary to permit continued operation of the facility? Is the work necessary or simply desirable? Will any significant benefit result from the work? What is the impact of not performing the requested work?

Value

After receiving a work request, the estimated cost of the work is determined. If the work is proposed as an improvement, the payback period should be considered. If major repairs are requested, a repair versus replacement study should be made. If the requested work is a modification subject to reimbursement, the initiator should be made aware of the estimated cost to verify that the work is still desired.

Urgency

The final factor evaluated for every work request is the relative urgency of the request. This urgency is established by the impact of the deficiency on the facility.

Projects are assigned priorities according to the urgency of need. If a deficiency prohibits operation of a facility, the costs can be staggering. Not opening a store, cancelling meetings, refunding tickets for a movie, or stopping the production line are all costly outcomes that might result from a major system failure. In such cases, immediate repairs are essential. The situation is an emergency. Other factors might less severely encumber the use of a facility and would have a lesser priority. Still other projects might have no significant impact but, being desirable, would still be done eventually. Most requesting parties desire that their request be fulfilled immediately. This is not usually possible; therefore, the various projects must be assigned a priority according to urgency of need. This process is described in more detail in the following section.

Project Priorities

Although it is theoretically possible to assign a definitive priority to each and every project, rank ordering the entire work backlog, such detail is not usually necessary. For a small maintenance work force, there is no need to formally classify the various work requests by priority. The maintenance manager would simply assign the tasks as deemed appropriate. In larger organizations, however, where several foremen or shop leaders schedule the tasks within their work element, the assigning of a priority classification is essential. Projects are generally assigned one of four or more general *priority classifications*. A project is assigned a priority to suggest the desired order in which projects should be completed. Assigning priorities assists the foreman in scheduling and also keeps the maintenance effort consistent. For medium- to large-sized maintenance organizations it is generally sufficient to classify projects in these four priority groupings: *emergency, urgent, routine,* and *deferred.* Using these groupings, the individual evaluating work requests assigns priorities according to the desired time frame of the repair.

Emergency Priority

The emergency priority is reserved for those projects which truly stop the use of a facility. Work on emergency priority requests should commence immediately and continue until the facility is restored to sufficient use. For example, water flow from a broken water pipe must be stopped immediately. This example demonstrates the nature and limit of most emergency priority projects; the object is to arrest and control the immediate problem. Once the emergency is abated, more information is gathered and a repair method determined. That repair is classified as a separate project, evaluated, and assigned a separate priority rating. If the pipe which was leaking fed only the lawn sprinkler system, stopping the leak is an emergency while repairing the pipe might be performed at a later time. A criteria for emergency priority work might be:

> The response should be made within 15 minutes of notification of the problem and work shall continue as rapidly as possible until the emergency is abated.

Urgent Priority

One step lower than emergency, an urgent priority is assigned to those projects which, while not completely prohibiting use of the facility, represent a threat to full facility use. The urgent project is normally started on the day it is reported. Urgent priority projects include threats to the safety of the building occupants. Examples are a malfunctioning fire alarm or sprinkler system. In such projects, there is usually no visible or physical restriction placed on building use, but continued exposure of the occupants is an unacceptable risk.

Projects which physically diminish the usability of the facility are also classified as urgent. If a single light fixture over a production line went out, for example, the workers might be inconvenienced, but not endangered. An urgent priority might be assigned, however, if the lack of full lighting could potentially lead to errors in some critical task. An urgent priority task should generally commence within four to six hours of notice and proceed until completed.

Routine Priority

A routine priority rating is assigned to the majority of the work requests received. A single burned out light bulb or a dripping faucet generally may be remedied within five to seven days. Any work must be completed eventually, but does not limit or threaten facility use or occupants, is given a routine priority.

Deferred Priority

The final category of priority ratings is deferred priority, used for those projects which are not necessarily required but are desirable. If no objective value can be assessed for a project it is usually classified as deferred. As a general rule, work should commence on deferred projects within thirty days of receipt unless seasonal or other considerations allow or dictate a greater delay in starting. Painting an office, replacing ceiling tiles, or replacing carpets are typical projects for which immediate completion is not essential but nevertheless must be completed at some time.

Advanced Systems for Work Priorities

The four classifications of work priorities previously described are sufficient for most medium-sized organizations. The moderate amount of identified work results in small numbers of projects within each of the priority categories. The shop foreman is able to reasonably schedule the work from these four defined groups.

In very large organizations, the sheer number of maintenance projects evaluated as routine would force the shop foreman to judge the relative priority of the many routine projects. However, when the number of projects within a single priority category far exceeds the capacity of the work force, the decision on which project is to be executed first should be made by the maintenance management staff, not left to the shop foreman. In these cases, more priority groupings are necessary to convey the desired order of execution of work to the shop foreman. Shown in Figure 7.1 is a sample list of ten project priorities. The priority is expressed as a number. The highest priority is "1".

These priority groupings better inform the shop foreman of the order in which to execute projects. When all priority "1" jobs are complete, then priority "2" work commences. This process continues until the project list is exhausted or the shop runs out of available workers.

There are numerous other ranking systems involving the assignment of priorities to the projects. One popular method employs the weighted scoring of the project by attributes. The maintenance manager determines which attributes will be evaluated and then determines the relative importance of each of these attributes. Typical attributes which might be chosen are:

- effect on facility user safety,
- effect on facility operations,
- benefits derived from the project, and
- cost of the project.

These attributes are assigned weights of, for example, 20%, 40%, 25%, and 15%, respectively. A project is then "scored" on a ten to one basis for each attribute; that score is multiplied by the weight and the resulting total is the overall priority score. The higher the score, the higher the priority and the sooner that project should be executed. When using this method, the scoring must be consistent. An expensive project which yields high benefits would receive a "10" in benefits, but a "1" in cost. Figure 7.2 is a sample scoring for a burst water main. Figure 7.3 is a sample scoring of an improvement project with a two year simple pay-back. Figure 7.4 is the scoring for the replacement of a faulty smoke detector.

When this scoring method is employed, the maintenance manager should check the relative weighting carefully. Poorly assigned weights will yield faulty total scores. This is a sensitivity analysis which is crucial if it is the only criteria used to establish priorities. Once the weights and resulting priority rankings have been validated, the shop foreman schedules the projects according to rank.

The Formal Evaluation Process

Most projects are evaluated for propriety, need, value, and urgency, followed by the assigning of a formal priority. The actual order in which these factors are assessed varies depending on the character of the work required. There is no one single individual who makes the evaluation decision. Each person who becomes aware of the project makes judgments concerning its path. For the sake of simplicity, however, the total maintenance staff can be placed into three groups to define when and who makes decisions. The work receptionist (also called the trouble

Priority	Project Description
1	A system, equipment, or component failure which jeopardizes the life safety of facility occupants.
2	A system, equipment, or component failure which prohibits the use of an important portion of the facility.
3	A system, equipment, or component failure which completely prohibits the use of the entire facility.
4	Problems which, if left unattended, will render all or a portion of the facility unfit for use.
5	Problems which decrease the productivity of facility users or increase facility operating costs.
6	Routine minor problems which, if not corrected, will eventually lead to more costly repairs.
7	Improvement projects.
8	Modification projects.
9	Replacement projects.
10	Projects which, while desirable, provide little definitive benefit for facility operation.

Figure 7.1

desk) is the first group. As the person who first receives notice of a problem, that individual determines whether the problem presented is an emergency situation requiring immediate action. The second group, the maintenance management staff, consists of the maintenance supervisor, the planner/estimator, and the scheduler. The final group is the maintenance execution work force, consisting of the shop foremen and their respective craftsmen. Figures 7.5 through 7.8 are flow charts tracing the roles of the three groups in the evaluation of various types of maintenance projects.

Scheduling Maintenance Activities

Two elements are scheduled in the course of maintaining facilities: the work activities and the workers who perform them. A well coordinated schedule of both work and workers may help minimize maintenance costs. Most maintenance workers are full-time employees with different skills and specialties. The individual scheduling the work must consider all available resources when developing the daily or weekly work assignments.

The first step in developing work assignments is to establish those work items which must be completed in the upcoming period. This work includes:

- any emergency or urgent priority work;
- recurring work normally conducted during the period;
- work underway, not completed during the prior period;
- mandatory preventive maintenance items;
- work which must be coordinated with the facility users;
- work which must be coordinated with the facility operation; and
- daily scheduled housekeeping type work.

Weighted Scoring Method for Project Priorities

Project Description: Burst Water Main

Attribute	Weight	Project Score	Extension
Effect on facility operations	.4	10	4
Effect on facility user safety	.2	5	1
Benefits derived	.25	10	2.5
Project cost	.15	1	1.5
		Total	9

Figure 7.2

Weighted Scoring Method for Project Priorities

**Project Description: Install Building Insulation
(Project has 2 years simple payback)**

Attribute	Weight	Project Score	Extension
Effect on facility operations	.4	1	.4
Effect on facility user safety	.2	1	.2
Benefits derived	.25	10	2.5
Project cost	.15	4	.6
		Total	3.7

Figure 7.3

Weighted Scoring Method for Project Priorities

Project Description: Replace Faulty Smoke Detector

Attribute	Weight	Project Score	Extension
Effect on facility operations	.4	5	2
Effect on facility user safety	.2	10	2
Benefits derived	.25	6	1.5
Project cost	.15	2	3
		Total	8.5

Figure 7.4

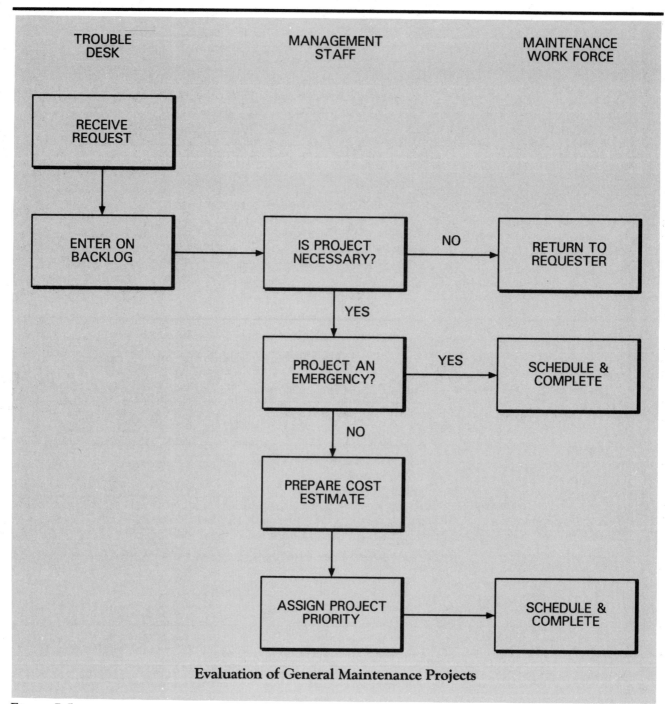

Evaluation of General Maintenance Projects

Figure 7.5

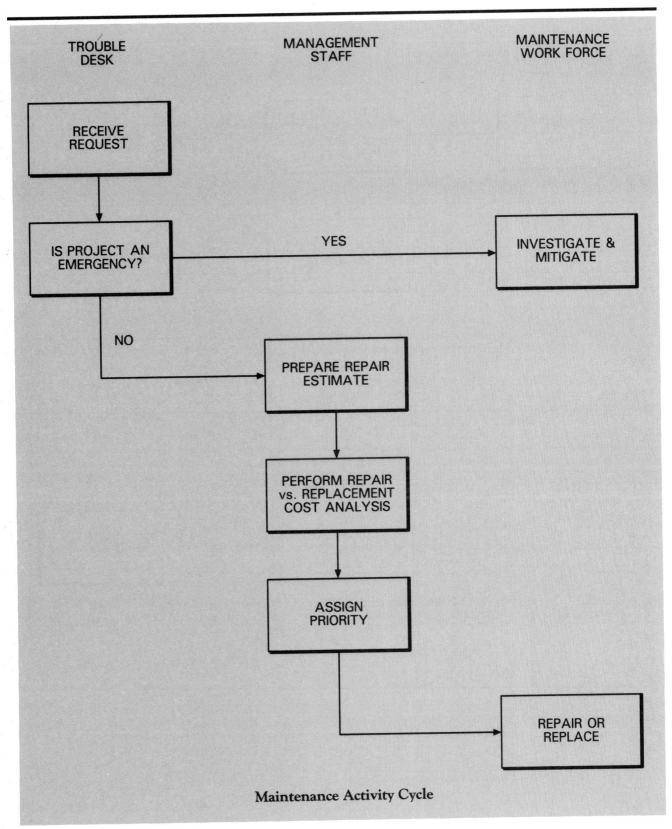

Maintenance Activity Cycle

Figure 7.6

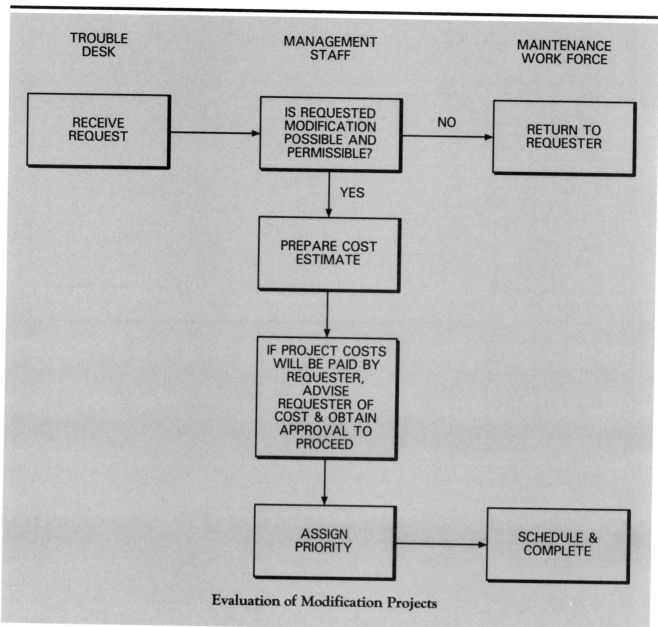

Evaluation of Modification Projects

Figure 7.7

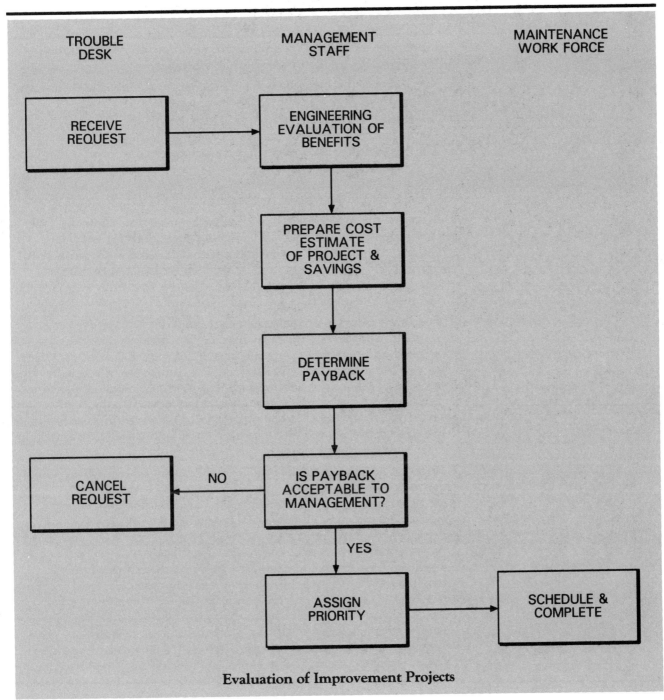

Evaluation of Improvement Projects

Figure 7.8

Having identified these items for the upcoming work period, the scheduler then assigns the available workers to the tasks. This action may result in the assignment of a full day of work for some workers, but, most likely, the work force is not fully scheduled.

The second step in developing the schedule is to identify and then assign workers to work items of the following categories:

- routine priority work;
- desired preventive maintenance work;
- optional preventive maintenance work deferred from an earlier work period;
- improvement projects;
- reimbursable modification projects.

Generally, the work force is completely assigned before all work in the above categories is exhausted. If, however, all workers are not fully assigned, their schedules are filled out with deferred priority work items. Only on very rare occasions should there be excess labor resources available when all work has been assigned. If this frequently happens, the maintenance work force is too large.

There are numerous algorithms similar to that described above for assigning tasks to workers. Many have been written as computer programs and purport to optimally match the work and workers. Unless the computer program is custom written to completely reflect the relative priorities of the maintenance manager and the specific skill levels of the members of the work force, it is not likely to replace a competent supervisor. A computer can be of great value, however, in balancing the workload between days. (See Chapter 9 for a more detailed discussion on the use of computer systems.)

Since it is not usually feasible to hire and fire maintenance workers on a day-to-day, as-needed basis, it makes sense to divide the planned workload evenly over the normal work week. This leveling is easily facilitated for planned or preventive maintenance work activities. If a particular batch of similar types of preventive maintenance is performed over the entire facility once each month, this batch can be spread out and scheduled for different weeks or days based on the estimated time required for their completion. In developing this schedule, the maintenance manager or planner/estimator can also consider the geographic location of the work to minimize transit time between jobs. Further, the maintenance manager can establish the schedule to accommodate noted cycles in repair work. If the maintenance manager observes that a large amount of repair work is noted and reported on one day of the week, the planned portion of the workload can be minimized for that day. The process by which work orders are scheduled and workers are assigned is shown in Figure 7.9.

Directing the Work

Once identified, prioritized, and scheduled, the work order must be communicated to the work force; a worker must be assigned and sent out to perform the work. The description of the desired work in the form of a work order must be physically transmitted to the shop supervisor or foreman. Once received, the foreman must direct a worker to perform the work. Following the assignment of the work, the foreman must have the capability to contact the workers for emergency repair work. This process is described in the following section.

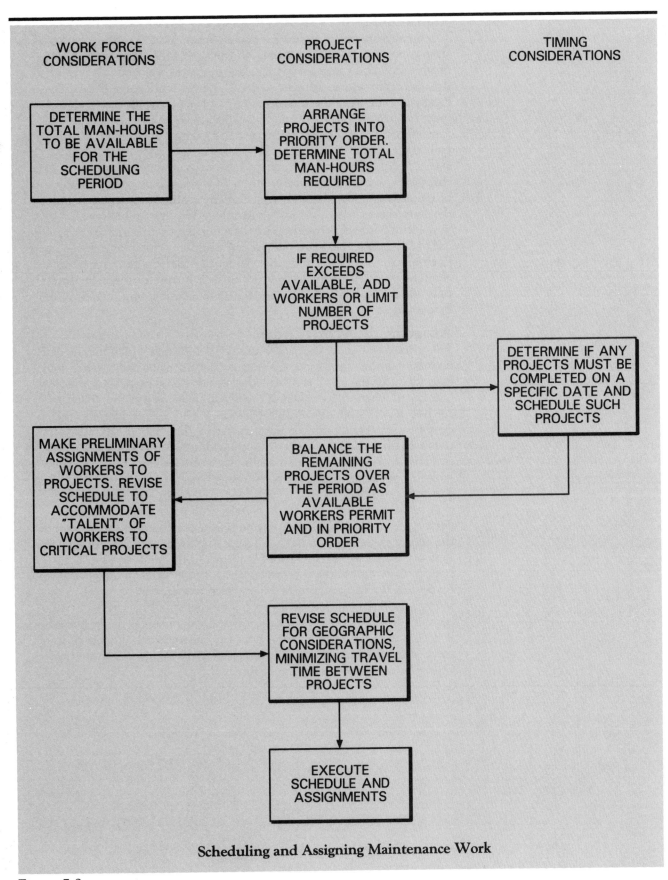

| WORK FORCE CONSIDERATIONS | PROJECT CONSIDERATIONS | TIMING CONSIDERATIONS |

DETERMINE THE TOTAL MAN-HOURS TO BE AVAILABLE FOR THE SCHEDULING PERIOD

ARRANGE PROJECTS INTO PRIORITY ORDER. DETERMINE TOTAL MAN-HOURS REQUIRED

IF REQUIRED EXCEEDS AVAILABLE, ADD WORKERS OR LIMIT NUMBER OF PROJECTS

DETERMINE IF ANY PROJECTS MUST BE COMPLETED ON A SPECIFIC DATE AND SCHEDULE SUCH PROJECTS

MAKE PRELIMINARY ASSIGNMENTS OF WORKERS TO PROJECTS. REVISE SCHEDULE TO ACCOMMODATE "TALENT" OF WORKERS TO CRITICAL PROJECTS

BALANCE THE REMAINING PROJECTS OVER THE PERIOD AS AVAILABLE WORKERS PERMIT AND IN PRIORITY ORDER

REVISE SCHEDULE FOR GEOGRAPHIC CONSIDERATIONS, MINIMIZING TRAVEL TIME BETWEEN PROJECTS

EXECUTE SCHEDULE AND ASSIGNMENTS

Scheduling and Assigning Maintenance Work

Figure 7.9

171

Transmitting the Work Order

The manner in which work orders are communicated to the foreman or shop supervisor varies depending on the size of the maintenance organization and the geographic location of the foreman with respect to the maintenance management staff. Where the work force is in a remote location, away from the maintenance management staff, the work order must be delivered to that remote location. Several options are used for transmittal of routine work orders. Most commonly, the work orders are picked up by the shop personnel or a messenger once or twice daily. This is functional, but requires frequent reliance on the telephone for emergency work. Where the quantity of work allows, the computer can be employed to electronically transmit (by a print-out) the work order to the shop supervisor or foreman's location. This method allows for immediate hard copy transmission of emergency work and minimizes the use of the phone. However, unless the computer system employed has an internal means for indicating that the work order has been received, such assurance is often obtained by a phone call. If the electronic transmittal must always be confirmed by phone, there is little point in transmitting by computer.

Assigning the Work Order

The foreman or shop supervisor usually determines which member or members of the work force are assigned each specific task. Actual assignment of the work is done either in person or by posting the work to a board where the workers can pick up their assignments. Each method has its strengths and drawbacks. When the foreman has the opportunity to talk with the workers before they depart on a work order, the foreman can impart any particular additional directions and precautions beyond those printed on the work order. The act of talking with the workers, while time consuming, does serve to lessen the chance that work will be forgotten or will be done improperly. There are drawbacks, however. The workers may become dependent on this personal contact and, should they complete their assigned work before the end of the work day, have to locate the foreman to seek more work. Less motivated workers will simply return to the shop and be nonproductive until the foreman returns. If the foreman directs the work by posting it to an assignment board, each worker can be given, at the start of the day, sufficient work to ensure their full-time productivity. A combination of the two methods of work assignment is the best solution. It is helpful to have the personal contact, but there should also be a method for the workers to pick up additional assignments should work orders be completed faster than estimated.

Communications

The maintenance manager must often immediately contact the shop foreman or an individual worker for an emergency repair. If a facility has and routinely uses a public address system, that is the resource to employ. In large facilities, however, workers may be dispersed over several buildings or out of doors, and a telephone or public address system is not feasible. In such cases, a paging system is generally used. The current advances in technology have made the use of a paging system very inexpensive and the systems have become more sophisticated, allowing short messages to be printed on a small display on the pager. Radios can be used where the sudden receipt of a message would not disrupt the worker or the workplace. Where the use of a radio would be disruptive, an earphone can be used.

The method of communications between the maintenance management staff and the shop foreman, and that between the shop foreman and the workers, is determined by the need for such contact. Where it is essential that personnel be reached immediately and it is not feasible to have each party near a phone at all times, the radio or "beeper" is necessary. For very large organizations it is not usually necessary for all parties to be "reachable" at all times. In such cases, the link between the maintenance management staff and the foreman is most critical. The foreman has many craftsmen engaged in routine scheduled tasks and will only need immediate communication capability with those few key craftsmen needed for emergency responses. Figure 7.10 is a tabular summary of recommended communication links between management, shop foremen, and craftsmen for small, intermediate, and large maintenance organizations.

Summary

The maintenance manager should evaluate each project after a work request is received and before work commences. This evaluation results in approval or disapproval of the project and, if approved, a priority is assigned to each one. The work request has now become a work order. That order is conveyed to the shop foreman who, in turn, schedules the work and assigns the workers. The evaluation of a work request offers the greatest potential savings in maintenance cost as the request can be denied or the scope of work redefined.

Communications Methods for Maintenance

Legend: Direct—Two way, face to face communication
          Pager—One way from sender to receiver, same as public address system use
          Radio—Two way, instantaneous
          Phone—Two way, potential delay
          Drop-Off—One way, left for later pickup, 4–6 hour delay
          Mail—One way, 24–48 hour delay

Type of Message to be Passed	From	To	Preferred Method	Alternate Method
SMALL MAINTENANCE ORGANIZATIONS				
Routine work request	Work receptionist	Maintenance Man	Direct	Drop-Off
Emergency work request	Work receptionist	Maintenance Man	Pager	Direct
INTERMEDIATE MAINTENANCE ORGANIZATIONS				
Routine work request	Work Receptionist	Planner/Estimator	Drop-Off	Direct
Routine work request	Planner/Estimator	Shop Foreman	Drop-Off	Mail
Routine work request	Shop Foreman	Craftsman	Direct	Drop-Off
Routine work request	Craftsman	Shop Foreman	Drop-Off	Direct
Preventive maint. order	Work Receptionist	Shop Foreman	Drop-Off	Mail
Preventive maint. order	Shop Foreman	Craftsman	Direct	Drop-Off
Emergency work request	Work Receptionist	Shop Foreman	Radio	Phone
Emergency work request	Shop Foreman	Craftsman	Pager	Radio

LARGE MAINTENANCE ORGANIZATIONS

Generally, large maintenance organizations will follow the same pattern for communications as intermediate-sized organizations. The primary difference, if any, will be that there are sufficient craftsmen within each shop that only a limited number might be involved in emergency repairs. In the intermediate organization, all craftsmen, not reachable by a public address system will carry beepers or radios, since they are likely to be needed for any emergency.

Figure 7.10

Chapter Eight

CONTROLLING THE MAINTENANCE EFFORT

Chapter Eight

CONTROLLING THE MAINTENANCE EFFORT

There is much work involved in identifying, evaluating, scheduling, and assigning maintenance work. Once the work has commenced, however, maintenance management responsibility continues; the work must be monitored to ensure proper completion. The results of each project must be compared to that of previously completed projects to determine the efficiency of the maintenance program. Trends in manpower utilization and timeliness of work should be monitored to determine if there is a need for enhancement in the number or the mix of the maintenance staff. The use of labor, materials, and tools must be controlled to minimize maintenance costs. This chapter contains descriptions of methods which may be utilized to control the maintenance effort.

Monitoring Maintenance Work

Every member in an organization, from the facility user to the maintenance foremen, has an interest in the progress of each maintenance work activity. The *facility user* is first interested in knowing that the request for work has been received and not lost or overlooked. This worry can be alleviated by returning to the user, an approved (or denied) copy of any submitted request. This return copy should contain some indication of the priority assigned by the maintenance manager. It should also include the *maintenance work request control number*, should the user desire additional information at a later date concerning the request. Once the user knows that the work request has been received and approved they will wait for completion of the work. Users are generally patient up to a point. For this reason, it is desirable to keep the users periodically advised of the current status of any outstanding work orders for their facility.

The *foreman* or shop supervisor should continually monitor all maintenance work, in order to ascertain:

- that the quality of workmanship is acceptable;
- that the assigned work crew is appropriate;
- that the rate of production is satisfactory; and
- that the work is completed and the work site left clean.

This monitoring is accomplished through periodic inspections performed either during or after the completion of the work order. Because inspection of all work is usually not feasible, the foreman must rely upon the report returned by the workers. The workers should note the following information on every work order:

- the condition found;
- the work performed;
- the type and quantity of materials used;
- the man-hours expended;
- any deficient condition discovered which could not be corrected; and,
- the name of the worker or workers who performed the work.

The foreman should review all returned work orders at the end of each day to determine if additional work orders should be generated.

Maintenance work should be closely monitored in order to detect any need for changing the estimated time for completion of a project. Occasionally, when undertaking what was initially diagnosed as a simple repair, the worker discovers a much larger problem requiring greater time and materials than originally estimated. For example, a leaking valve, thought to need only a tightening of the stem packing gland, turns out to have a cracked casing requiring full replacement of the valve. Or, the leaking valve may require simple tightening of the packing gland. If the actual man-hours expended will be compared to the estimated man-hours as a measure of the worker's productivity it is essential that the estimates be revised when the scope of the work changes.

Maintenance work is also monitored in order to determine changes in the facility and its attendant equipment. The foreman should review all work orders to determine if there is a trend developing in the maintenance of any particular building or type of equipment. Repetitive repairs to faucets in a building could be an indication that the faucet washers, all installed at the original construction, have all reached the end of their useful life. In that situation the foreman would initiate a work order to replace all washers, thereby saving the continual transit time for repairs as the failures progress.

The final step in monitoring the proper completion of work orders is to have the user inspect the completed work and sign the work order. This step requires some additional time while the worker locates the user, but provides verification that the job was completed.

Analyzing Maintenance Work

The final facet in the management of maintenance activities is the analysis of the results of maintenance work to determine if the work is being accomplished at the lowest cost. It is in this area that a computer system is most helpful. For this reason, while developing the computer programs, the data to be analyzed must be identified. Then, the maintenance manager must ensure that this data is recorded. From data analysis, trends and potential improvements may be discovered in staffing, productivity, and the physical condition of the facility. These trends are most easily visible in reports generated from computer-stored data. (See Chapter 9 for more information on computer maintenance management systems.)

Manpower Utilization

The data collected from all work orders includes the number of man-hours that each specific trade used to complete the maintenance tasks. From this information, several comparisons can be made to determine the relative efficiency of current operations. Reports should be available to make the following determinations:

Reports of Total Shop Activity

- Breakdown of man-hours by work category
- Average time to complete work orders by priority
- Average man-hours available for unplanned work
- Average backlog of work remaining
- Quantity of preventive maintenance performed on time
- Average amount of time overdue on deferred work
- Amount and frequency of overtime
- Repair to preventive maintenance ratio

Reports of Individual Worker Activity

- Attendance
- Average actual man-hours vs. estimated man-hours
- Utilization by work category
- Amount and frequency of overtime
- Work breakdown by percentage

Work Breakdown by Percentage

Figure 8.1 is a sample report of man-hour utilization for a one-week period by four shops in a large maintenance organization. The work has been summarized by work categories and the percentages calculated for each category. The percentage breakdown for each shop is relatively predictable. Obviously the janitorial work force and grounds maintenance shop are primarily employed in housekeeping and general maintenance work activities. Therefore, shops which deal with sophisticated equipment can be expected to have a fair amount of preventive maintenance. Carpentry and sheet metal shops likely perform a high percentage of modification and improvement work. Such shops deal with large quantities of consumables and, therefore, make up a high percentage of the ordering and inventory of supplies, materials, and parts.

Workload breakdown for the plumbing and heating shops have a direct relationship to the condition of the equipment they maintain and the state of the preventive maintenance program. Although the work performed by each shop is sufficiently different to justify separate organizations, the character and quantity of the work for the plumbing and heating shops is similar. For this reason, the two should have a similar quantity of failures and subsequent repairs and preventive maintenance procedures. In Figure 8.1 the plumbing shop spends a greater amount of time on preventive maintenance than on repairs, while the opposite is true for the heating shop. Without complete knowledge of the facility, no immediate conclusion can be made from this observation. However, the large quantity of time spent on repairs in the heating shop may indicate one or more of the following problems:

- Preventive maintenance work is being ignored.
- The preventive maintenance program for heating equipment is deficient in content and/or frequency.
- The heating equipment has reached the end of its useful life and requires replacement.
- The heating shop may need engineering troubleshooting assistance.

SUMMARY OF MAN-HOUR UTILIZATION BY SHOP FOR THE WEEK ENDING

6/20/87

SHOP/ MAN-HRS/ PERCENT	HOUSE-KEEPING	GENERAL MAINT.	REPAIR	PREVENT. MAINT.	REPLACE-MENT	MODIFI-CATION	IMPROVE-MENT	SUPER-VISION	INSPEC-TION	ORDERING INVENTORY	TOTAL WORKED	VACATION	GRAND TOTAL
JANITORIAL													
MAN-HOURS	240.0	100.0	30.0	0.0	0.0	0.0	0.0	30.0	20.0	20.0	444.0	0.0	440
PERCENT	54.5	22.7	6.8	0.0	0.0	0.0	0.0	6.8	4.6	4.6	100.0		
PLUMBING													
MAN-HOURS	10.0	30.0	125.0	200.0	18.0	27.0	30.0	40.0	30.0	15.0	525.0	35.0	560
PERCENT	2.0	5.7	23.8	38.1	3.4	5.1	5.7	7.6	5.7	2.9	100.0		
HEATING													
MAN-HOURS	30.0	30.0	250.0	127.0	41.0	10.0	0.0	35.0	9.0	38.0	560.0	40.0	600
PERCENT	5.3	5.3	44.6	22.6	7.3	1.8	0.0	6.3	1.6	6.3	100.0		
GROUNDS MAINTENANCE													
MAN-HOURS	60.0	135.0	15.0	10.0	0.0	0.0	0.0	10.0	0.0	10.0	240.0	0.0	240
PERCENT	25.0	56.3	6.4	4.1	0.0	0.0	0.0	4.1	0.0	4.1	100.0		

Figure 8.1

- This week is an anomaly due to premature failure of some heating equipment due to extraordinary causes.

The last cause is generally predictable; a major repair takes a considerable amount of time and the maintenance manager might be pleased to know that at least some time was spent on preventive maintenance. The object of the report, in any case, is to give the maintenance manager an accurate picture of the past week's performance.

Long-Term Maintenance Trends

Subtle changes may not be noticeable on an individual weekly report, but the accumulated effect of these subtle changes becomes evident over an extended period. Figure 8.2 shows eight weeks of man-hour utilization for a plumbing shop. By quickly scanning each work category, it is obvious that the work in each category has remained relatively constant with no particular increase or decrease. One point worthy of note, and fully expectable, is the quantity of vacation time taken during the week ending November 27, 1987, due to the Thanksgiving holiday and the fact that many workers also take the following Friday off. A continual decrease in man-hours, however, is reflected in a decreased percentage of preventive maintenance tasks done on time. This is likely to lead to increased failures if not corrected in the following week(s).

Shop Backlogs

The backlog of work for a particular shop or work center is made up of the man-hours necessary to complete all outstanding unplanned work. This includes repair, replacement, modification, and improvement projects. It is inevitably likely that there are times when the backlog exceeds the number of man-hours available in one week. In Figure 8.2, the eight-week shop summary, the average number of man-hours available for unplanned work, is 192.1. This average is generated from the eight weeks of data, which included the "short" week containing the Thanksgiving holiday. The resultant average may be slightly lower than an average of non-holiday weeks, but should be used since over the year, most eight-week periods contain one or more holidays.

The calculated backlog alone indicates nothing about the maintenance organization or individual shop. A highly efficient, properly manned shop might still have a large backlog due to recently approved modification or improvement work orders. However, it is useful to calculate the backlog because long-term trends should be observed, which may help reduce maintenance costs. A slowly increasing backlog may indicate understaffing and signal the need for overtime or hiring additional workers. A stable, but high backlog might indicate the need to employ a mixture of overtime and temporary employees for a short period until the backlog is brought down to reasonable levels.

There are no definitive judgments which can be drawn from any single observed trend. Figure 8.3 is a tabular breakdown of the possible problem areas associated with a wide variety of observed trends. The recommended management action to reverse or stabilize the trend is also shown.

SUMMARY OF MAN-HOUR UTILIZATION FOR PLUMBING SHOP

EIGHT WEEK PERIOD ENDING 12/04/87

WEEK ENDING	HOUSE-KEEPING	GENERAL MAINT.	REPAIR	PREVENT. MAINT.	REPLACE-MENT	MODIFI-CATION	IMPROVE-MENT	SUPER-VISION	INSPEC-TION	ORDERING INVENTORY	TOTAL WORKED	VACATION	GRAND TOTAL
10/16/87	10	32	150	200	0	24	40	40	28	20	544	16	560
10/23/87	20	18	144	186	32	48	18	40	32	24	522	38	560
10/30/87	12	30	130	192	0	48	20	40	24	18	514	46	560
11/06/87	16	32	115	210	20	24	10	40	32	16	515	45	560
11/13/87	10	30	125	200	18	27	30	40	30	15	525	35	560
11/20/87	12	30	150	180	0	40	10	40	35	17	514	46	560
11/27/87	10	20	90	160	0	25	10	32	26	10	383	177	560
12/04/87	15	35	120	190	25	24	30	40	24	20	523	37	560
TOTAL	105	227	1024	1518	95	260	168	312	231	140	4040	440	4480
AVERAGE	13.1	28.4	128	189.7	10.6	32.5	21	39	28.9	17.5	505	55	560

AVERAGE TIME AVAILABLE EACH WEEK FOR UNPLANNED WORK: 192.1 MAN-HOURS
(TIME SPENT ON REPAIR, REPLACEMENT, MODIFICATION, IMPROVEMENT)

AVERAGE TIME TO COMPLETE WORK ORDERS:

EMERGENCY   - 0.7 DAYS
PRIORITY 1 - 1.1 DAYS
PRIORITY 2 - 3.6 DAYS
PRIORITY 3 - 8.7 DAYS

PREVENTIVE MAINTENANCE WORK COMPLETION RECORD:

ON TIME   - 78%
WITHIN 7 DAYS - 14%
DEFERRED   - 8%

REPAIR TO PREVENTIVE MAINTENANCE RATIO: .677

Figure 8.2

Trends and Indicators of Maintenance Program Deficiencies

Observed Trend	Possible Deficiencies	Solution
Increasing shop repair man-hours	Poor PM Activities. PM not being done. Equipment nearing age for replacement.	Review PM content & frequency. Review PM records. Schedule replacement.
Increasing shop vacation man-hours	Leave abuse. May be normal, seasonal.	Review leave policy. Check past records.
Increasing man-hours for housekeeping	Poor shop priorities.	Review staffing for possible reductions.
Increasing man-hours for supervision	Poor shop foreman priorities. Over delegation of duties.	Review policy with foreman. Review staffing level for possible reduction.
Decreasing man-hours for preventive maintenance	Deferred PM. Potential premature failures.	Examine shop foreman procedures for project priorities.
Any general increase in man-hours for any activity type	Potential drop in productivity. Poor supervision.	Inspect work quality. Observe worker habits. Query supervisor.
Any general decrease in man-hours for any activity type	Possible neglect of required work. Also possible increased productivity.	Ensure all work is being done. If so, reward/acknowledge workers and foreman.
Identical hours in one category for several consecutive weeks	Foreman probably using old data, not tracking actual man-hours.	Investigate, advise foreman. Discipline.
Increase in any shop backlog for extended # of weeks	Understaffed shop.	Examine shop work closely to ensure priorities are proper. If so consider increase to staff or contract for some services.
Shop backlog not changing, but unacceptably high	No problem.	Consider short-term use of overtime to reduce backlog.
Single reported week with high repair man-hours and low man-hours in all other categories	Probable that a major equipment failure required diversion of all personnel to that repair work.	Should not be a surprise. Major failures are usually remembered! But should check to ensure mandatory PM is performed.

Figure 8.3

Trends and Indicators of Maintenance Program Deficiencies (continued)

Observed Trend	Possible Deficiencies	Solution
Increasing time necessary to complete high priority	Not making prompt response. Possible weak communications between trouble desk and shop.	Verify method for prompt notification to shop of high priority projects Review policy with foreman.
Shorter average time to complete low-priority projects than high-priority projects	Shop foreman assigning wrong projects first.	Review policies with foreman.
Increasing complaints from customers about work quality	Poor inspection. Not getting customer signature on completed work requests.	Increase foreman inspection. Verify customer signatures.
Customer complaints about time needed to complete any work	May have misread the customers urgency of need. May be that organization needs and is willing to pay for a larger maintenance staff and faster response	Verify situation for specific projects. Advise upper management of problem, and seek increase to maintenance budget.
No customer complaints	No problem.	Such a rare event that it is probable the maintenance organization is overstaffed!

Figure 8.3 (continued)

Material Cost Analysis

Although costs of the labor necessary to perform maintenance tasks are widely variable, these costs are also controllable. The purchase of materials is unavoidable; the extent of the materials is dictated by the situation; and the only means for reducing the material cost is through rigorous price comparisons by the purchasing department. Worse yet, the need for immediate repair often reduces that search to one of immediate availability with little concern for cost. Accordingly, while material costs account for a significant portion of the total maintenance budget, the reduction of these costs does not deserve nearly as much attention as do labor costs. Reports on material cost should be developed using the same format as that previously shown for man-hour expenditure. The data for the reports is gathered at the shop level. The accuracy of the data placed on the completed work order is determined by the policy of the maintenance manager. If the shop has strict inventory control over all maintenance materials, it is likely that the costs attributed to each work order can be very accurately determined. In other cases, where actual material costs are not well known at the shop level, the costs of the materials are estimated by the workers or foreman. In either case, all maintenance costs should be measured against two budgets. First, the *material costs* for each individual work order should be compared with the *estimated costs* developed by the planner/estimator and significant variances investigated. A large increase of material costs over the original may indicate that the scope of the job has changed. Revision to the estimate would then be required. Or, the difference may indicate that the estimate was flawed due to reliance on out-of-date material prices or improperly determined material quantities. The *total material costs* for each week of work should also be compared to the *overall shop budget*. Concern should be given to long-term comparisons, although a large variance in an individual week should be noted. It may take several weeks or even months for a trend to appear. It may even be necessary to compare the totals from two widely separated, but equal, periods of time to identify an upward or downward trend in maintenance material costs.

A long-term increase in material costs may be traced to a change in work methods, such as when a foreman directs workers to replace parts rather than to attempt repairs. Similarly, a new inexperienced worker may resort to wasteful replacement due to an inability to perform repairs.

The possibility of materials being charged to a work order to facilitate theft in maintenance operations where strict inventory control is kept for all materials should not be overlooked. Certain trends can be accounted for and adjusted. For example, if the maintenance supervisor has noted that the shop foreman or workers consistently under or over report material costs, those costs must be adjusted to reflect actual expenditures. The greatest use for well-documented maintenance material costs is for future budget predictions. If the material costs are reported accurately, they can be used as a basis for budget development for future periods.

Historical Cost Data

Well-developed overall facility or system historical costs are extremely useful in budgeting for future periods or in predicting the maintenance costs of proposed new construction. Cost data is usually gathered by work center, shop, or trade. Budgets are developed in a similar manner. The total labor and materials costs for each separate building or section of a building is calculated from completed work orders. This is essential where the facility users are billed for all maintenance costs.

Past maintenance costs, by facility, facility component, or system, are particularly useful in detecting trends. Most frequently the trends show a slowly increasing maintenance rate as might be expected for an aging facility. Occasionally, however, the trend indicates the results of proper preventive maintenance with decreasing maintenance costs.

In-house developed maintenance costs should be provided to the architectural engineering design team to allow the incorporation of the least costly engineering or architectural systems. It is important to keep housekeeping, repair, and preventive maintenance costs separated from the costs of improvement and modifications. Since improvement or modification projects are executed to either accommodate a new function, incorporate new technology, or to correct an inefficient system, these costs are inappropriate when budgeting for future construction.

The presence of ever increasing maintenance costs, coupled with numerous proposed modification or improvement projects, may signal the need for a comprehensive rehabilitation of the facility. When a major increase in maintenance costs for a particular facility occurs, it is essential that the maintenance supervisor closely examine both the physical condition of the facility and the adequacy of the preventive maintenance program. Although systems in the facility will progressively degrade, it would be unusual to see a large increase in maintenance costs without some other factor being present. That factor may be just one large repair cost which masked an otherwise normal maintenance rate. Where the maintenance manager is responsible for several similar facilities, the maintenance costs of those facilities should be compared. This simple comparison is often the fastest possible way to highlight a change in the maintenance costs. If there is just one facility, a baseline should be developed over several months from which variances can be measured.

Equipment Histories

An accurate history of a particular piece of equipment is important for a foreman or craftsman to adequately and cost effectively perform maintenance on the component. By observing the nature, extent, and timing of the previous work done on a piece of equipment, the repairman can usually discern the probable cause of a malfunction. With an appropriate equipment history the worker can also go to the site prepared with the proper tools and parts. An equipment history should include the following data:

- Equipment designation
- Type of equipment
- Manufacturer
- Model number
- Location
  Building
  Room
  Area (above ceiling, etc.)
- Date installed
- Last overhaul
- PM history
  PM type
  PM frequency
  Last PM
  PM costs

- Parts required for PM
  Quantity
  Part number
  Description
  Cost
  Source of supply
- Major repairs
- Minor repairs
- Future maintenance
- Next overhaul date
- Next inspection
- Similar equipment elsewhere in the facility

This compilation of data is most easily tracked with a computerized maintenance program. A sample output of a computerized equipment history is shown as Figure 8.4. Without a computer, this system would have to be recorded manually. (For more information on computerized maintenance management, see Chapter 9.)

Supplies and Spare Parts

In order to be prepared for a prompt response to maintenance problems, the maintenance manager must ensure that sufficient materials are on hand. Methods for monitoring and controlling supplies and spare parts are described in the following section.

Consumables

Consumables are items which are routinely expended in the course of facility operation and maintenance. This category includes raw construction materials, tools, and maintenance equipment. Housekeeping activities are the primary users of consumables. From toilet paper to floor wax, these materials are essential to task completion. Within other maintenance activities, oil and grease used in preventive maintenance of fixed equipment are other consumables.

For consumable items, the amount of stock necessary for any period of time is easily determined, the need for the material is constant, and the alternatives are limited. The primary mechanism for reducing the cost of consumables lies in price comparisons by the purchasing staff.

Depending on available storage space, some quantity discounts can be obtained. Generally, however, the cost of constructing storage space or dedicating useful spaces to storage far exceeds any potential discounts which can be achieved by purchasing consumables in large quantities.

Construction Materials

The use of raw construction materials is most commonly associated with improvement and modification projects. The costs of these materials are considered in preparation of the estimated cost of the proposed project and are therefore not usually purchased as stock materials. Since the prices of various construction materials are relatively constant among suppliers, there is little reason for comparative shopping. The primary consideration in managing the purchase and consumption of construction materials is the high degree of waste generally associated with small projects. However, since construction materials are readily available, vary widely in type and size, and are bulky to store, there is no advantage to pre-purchasing bulk materials. Further, the degree of waste in use of these materials makes inventory control difficult and leaves these materials particularly vulnerable to theft.

```
EQUIPMENT HISTORY CARD                    |Equipment Control Number
                                          |    AC-02-07-006
                                          |
_____|_____

Equipment Name
       AIR HANDLING UNIT - 12,500 CFM

_____

Equipment Data
   Manufacturer                   |Model Number
      HATHAWAY MFG.                |     H-125
                                   |
_____
Equipment Location               |Specific Installation Area
   Building    |Room              |
    ATWOOD     |   473            |  SUSPENDED FROM OVERHEAD
               |                  |
               |                  |
_____
Date of Original     |Installed By              |Installation Cost
  Installation       |                          |
    12/63            |SOMERTON CONSTRUCTION      |  UNKNOWN
                     |                          |
                     |                          |
_____|_____|_____

PREVENTIVE MAINTENANCE ORDERS FOR THIS EQUIPMENT

_____
                 |                           |Date Last  |Date
PMO CONTROL #|Frequency|  Description         |Completed   |Next Due
   0207-01       W        INSPECTION            01/07/88    01/14/88
   0207-02       M        LUBRICATION           01/21/88    12/20/87
   0207-03       A        FILTER REPLACEMENT,
                          SERVICING             05/20/88    05/28/87



_____

Equipment Repair History
            |                                        |
Date        |   Description                          |Repair Costs
_____|_____|_____

 05/13/71      REPLACE BELT AND BEARINGS                 $150.00
 10/09/82      REWIND AND REBUILD MOTOR,
                  REPLACE BELT                           $477.00
```

Figure 8.4

Tools

The purchase of tools represents an opportunity to achieve some moderate increases in productivity and, hence, some savings in labor costs. The cost of purchasing the specific tool and converting to a new method of maintenance may outweigh the benefits to be derived. The most important factor to analyze when considering a tool which allows more efficient maintenance is not the *purchase itself*, but, rather, the *long-term benefit* derived from the use of that tool. While the suggestion to procure a labor-saving tool often comes directly from the workers, the maintenance manager should realize that the objectives of the workers to make the job easier and the objective of management to reduce maintenance costs are not always closely tied. Therefore, accompanying the purchase of a new tool must be a management and labor commitment to make the product work to best advantage.

Spare Parts

Spare parts are required on infrequent occasions. These parts are individual replacement components kept on hand to replace failed components of facilities or facility systems. While there are many valid reasons for retaining any particular spare part, the primary reasons are *predictability*, *vulnerability*, and *availability*. Many component parts of building equipment and systems are expected to wear out. Moving parts, parts exposed to corrosive agents, and parts which are exposed to extreme heat or cold degrade under use and are likely to require replacement. These parts are also usually integral to the proper operation of the system.

A particular spare part should be retained primarily to avoid lengthy down time of critical systems or facilities. When the continuous operation of a facility system is critical for life safety or profitability, that system must be protected by an accurate inventory of spare parts. Most often, such a critical system is first protected by some redundancy designed into the system. A motorized steam valve, for example, is likely to be installed with a manual bypass loop, allowing at least manual operation and modulation. Such a manual operation would not be feasible for extended periods, however, so a spare motorized valve should be kept in inventory. A spare parts inventory must be developed with a sense of balance, assessing the risk of failure and its attendant cost against the cost of procuring and storing the spare part in question.

Taken to the extreme, a system would have complete built-in redundancy. In this case, there would be an automatic switch over to the backup mode if the primary component fails. This redundant system is then backed up by a complete set of all replacement parts. Such extremes are, in fact, used by some facility systems. For example, electric utility companies develop multiple paths for providing electricity to each of its customers. By doing this, the utility can re-route power manually or automatically around failed wires. Further, the utility retains spare switches, wire, and transformers to facilitate prompt repair of the failed component while the redundant system continues to provide service.

The utility company's method of redundancy and spare parts inventory exemplifies the most efficient method of reducing spare parts costs, because the utility also maintains a firm policy of standardization of system components. This means that just one spare part can be used to replace parts in many individual sections of the electrical system. This policy should be extended to all building systems throughout a major facility. In addition, compatible types and models of components should be purchased when modification, improvement, or new construction projects are developed.

The cost of a spare part extends beyond the initial purchase price; the complete inventory must be stored in an accessible location. However, the cost of providing these storage spaces is not recovered through the productive use of the space. Therefore, the cost of constructing and maintaining these storage spaces must be divided proportionately amongst the various departments of the facility.

Spare parts should be retained according to the time required to acquire a replacement part. It is not unusual to have to wait months for a particular part. Down time of a few days may be fully acceptable for some systems, but a longer period is generally unacceptable. For this reason, the facility maintenance manager must assess the potential for failure of each particular component, determine the acceptable length of time for restoration of system operation should that component fail, and determine the lead time for purchasing a replacement. From this analysis, the facility maintenance manager decides whether it is necessary to pre-purchase that component. Manufacturers usually provide a recommended list of spare parts for each major piece of equipment. Therefore, an analysis of these parts is not necessary. If a particular part is very expensive, however, the maintenance manager should perform a detailed analysis to determine its material and supply needs. For less expensive parts, the cost of purchase and storage may be less than the cost of the manager's time spent investigating the situation. Figure 8.5 shows in chart form the decision process for developing a spare parts inventory.

Inventory Control

Accounting and scheduling form the basis of inventory control. An inventory control system must accurately keep track of each part through the steps of identification, ordering, receiving, storing, and expending the individual items. The sophistication of the inventory control methods depends on the size and importance of the inventory. If the maintenance work force is small, the facility systems simple, and the number of modification projects negligible, the spare parts inventory may be kept mentally by the maintenance foreman, observing the inventory on the shelves of the storage room and determining when stock is low. As the complexity of the facility and the maintenance organization grows, a system for monitoring the level of various consumables, construction materials, tools, and spare parts must be formalized. This can be done manually, but is more efficient when a computer is utilized. Manually, inventories are usually kept on stock cards in a large file. When an order is received, the quantity of new stock received is added to the stock card. When the materials are expended, the quantity used is subtracted from the card. If this system is faithfully updated and there is sufficient security on the stored materials, the card can be relied on to depict the state of the inventory. This manual method has some distinct drawbacks, however. First, the system is a tedious expenditure of time. The cards must be reviewed frequently and verified with an actual counting of materials. In addition, tracking parts which have been ordered, but not received, does not fit well with the card system. Unless the foreman is prepared to drop everything and prepare a replacement order each time a particular material or spare part is used (and the level of stock is below the desired on-hand level), the foreman will have to periodically review the entire set of inventory cards to prepare orders.

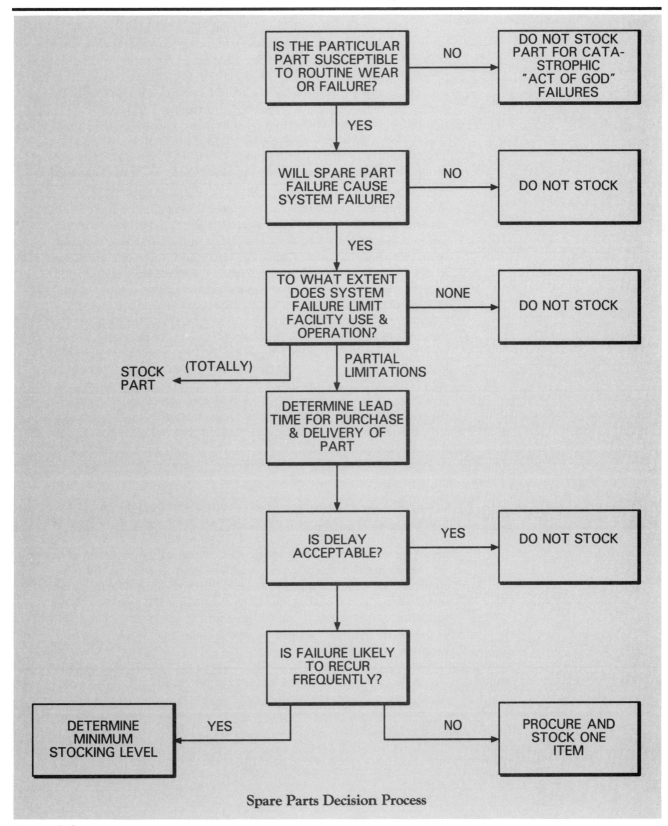

Spare Parts Decision Process

Figure 8.5

The second factor in managing an inventory is that the need for and consumption of many maintenance materials is highly predictable. This is particularly true for those materials utilized in planned work such as preventive maintenance tasks. (See Chapter 11 for more information on preventive maintenance.) The exact type and quantities of related material supplies is known, and the time it will be needed is predicted by the designated frequency of the task. For this reason, a total yearly requirement for preventive maintenance materials can be developed. This task is easiest if the preventive maintenance program is computerized. The computer inventory system may be set up to automatically generate orders for inventory items when the stock level reaches a designated low threshold level.

The final aspect of inventory control is to maintain accountability for materials and spare parts. The materials and spare parts used in the maintenance of commercial facilities are common to those materials and spare parts needed for private home repair and, therefore, are susceptible to pilferage. A formal system of inventory control should detect most occurrences of theft. Construction materials, such as lumber, pipe, piping fitting, nuts, bolts, and nails, which are readily usable for many applications are most likely to be stolen. Therefore, the storage of these materials must be secure. Additionally, since these materials can be purchased almost anywhere, the stock level for these items should be kept relatively low. Although there may be discount prices associated with bulk purchases, this may be more than offset by the increased possibility of loss to pilferage. Consumables such as toilet paper and paper towels are also susceptible to theft. Since this stock is usually stored throughout the facility, strict inventory control is not feasible. In this case, analysis of ordering levels must be performed continually to determine increases in consumption which may be attributed to theft.

Summary

Controlling the maintenance effort means controlling the costs of maintenance. The costs of maintenance are a function of the *quality* of work performed and the *quantity* of labor, materials, and equipment used. These factors must be monitored on individual projects, as well as for the overall maintenance program. The maintenance program is controlled by examining long-term trends for the expenditure of labor and materials and adjusting maintenance staffing and methods accordingly. Just as problems within a facility are identified and solved, problems within the execution of that maintenance work must be identified and solved.

Chapter Nine

COMPUTERIZED MAINTENANCE MANAGEMENT

Chapter Nine

COMPUTERIZED MAINTENANCE MANAGEMENT

Computerized facilities maintenance management systems, previously available only to very large facilities, are now accessible to small and medium-sized maintenance organizations. Advertisements for such products promise reductions in maintenance costs. Such claims are well founded. Where a manual system of maintenance management is currently employed, moderate reductions in maintenance costs and facility down-time are fully possible. Where no formal maintenance management procedures are present, the structure and ease of a computerized maintenance system can result in dramatic cost reductions and improvements in facility availability. There are, however, practical limits to the applicability of computerized maintenance systems. Very small organizations may not be able to justify the simplest and least expensive system. Intermediate-sized organizations may not be able to justify the cost of a very sophisticated computer system. Large maintenance organizations may find the effectiveness of their maintenance program encumbered by a system which has less than complete features. The task, then, is to match one of the many available computerized maintenance systems with the size and needs of a particular maintenance organization.

Reasons for Computerization

The objective of any maintenance program is to minimize the total costs resulting from the execution or lack of execution of proper facility maintenance. Since these costs generally accrue in small increments through the execution of many small maintenance efforts, the ability to track each of these activities and their attendant costs is of great importance. Once such data is gathered, it must be interpreted and appropriate actions taken. The computer, because of its ability to store and manipulate large amounts of data, can be a valuable asset to the facilities maintenance manager. A computer can quickly scan this data and report specific findings, trends, or discrepancies. The advantage of the computer over manual methods lies primarily in the ability to store, process, and report large volumes of various types of information.

Information Storage and Processing

For each individual maintenance activity, at least two pieces of information are necessary: *what* needs to be done and *when* it must be done. If resources to perform maintenance were not limited and if the facility could accommodate interruption at any time, the work list could be kept manually, sorted by "due date" and executed in order. Obviously maintenance is not that simple. Resources are limited, cost is important, and the maintenance effort must accommodate the continued operation of the facility. When these factors are considered, the volume of data related to each activity increases proportionally. Other factors include the location of the project and the shop responsible for the project. Each of these facts related to an individual project can be listed on a single sheet of paper and stored until the project is to be executed. The total maintenance backlog would then be a single pile of individual projects.

The different information related to an individual project has numerous applications. A list of all projects for a single shop is needed. A listing of all projects for a single building or facility function is also important. A summary of past costs sorted by building, shop, or type of work would be useful to analyze trends. This data can be used to organize and manage the maintenance effort. However, retrieval of this information requires a considerable amount of manual searching. A computer can save time during information storage, retrieval, sorting, and reporting. Once salient data concerning an individual project is stored within the computer, it can be manipulated and combined with other project data in endless combinations, saving time otherwise spent in manual compilation of data. Modern data storage devices, such as hard disks, can retain millions of bytes of data.

Automated Calculations

Most maintenance management actions follow a logical pattern; each project follows a series of common steps. A project is identified, evaluated, given a priority, scheduled for accomplishment, assigned to a craftsman or service contractor, completed, and inspected. Data from the project is recorded for historical records, including the estimated and actual costs of materials, equipment, and labor. Many of these steps may may be duplicated by a computer program, and the speed of calculation and data processing of a computer produces immediate results. Manual calculations are more time-consuming. There are many maintenance activities which are regularly scheduled, but only at great time intervals; the computer provides automatic recall of these requirements, reducing the possibility that such maintenance activities will be forgotten. However, the move from manual recordkeeping to computerized maintenance management is a costly one. In addition to the purchase cost of the computers and computer programs, the move to a computer system implies a long-term commitment of labor to enter and retrieve the data from the computer. Those costs must be compared to the anticipated benefits of a computer system before this major commitment is made.

Standard Program Features

The maintenance procedures described in this book represent the logical evolution of maintenance as a managed technical activity and are employed with minor variations in most well-run maintenance organizations. Because of this commonality of approach within the industry, computer programs, or software, have been developed which allow for immediate computerization of a maintenance program. These programs have the standard features which are outlined in Chapters 6, 7, and 8. The following sections briefly describe the basic facets of each standard feature. Following each description is a summary of the available enhancements to each feature. Not all commercially available programs contain every one of these enhancements; therefore, the facilities maintenance manager should examine each of these features and note the attendant enhancements. A listing of the essential features desired should be used as a checklist for the evaluation of the various commercial software packages.

Where applicable, computer printed forms and reports are shown as examples of computer possibilities. Most maintenance management software packages produce varying versions of these forms and reports.

Unplanned Work Processing

All maintenance programs must react to unforeseen maintenance activities, defined earlier in this book as unplanned work. The scope of the work and its location must be captured. As the project is evaluated, planned, scheduled, and assigned, more information is obtained. When the work is in progress and eventually completed, the amount of information grows. This information must be entered, transmitted to appropriate personnel, continually updated, and compiled to produce weekly, monthly, or annual reports.

Entry: All initial data must be entered manually into the computer. This is generally performed by the person manning the work reception desk as a phoned in request is made or as time permits for those requests submitted in writing. The specific form used for requesting work should be identical to that produced by the computer for phoned in requests, as the work request should be a constant thread throughout the processing of maintenance work. The work request is used for describing the work to the worker, who records on it the maintenance results. Figure 9.1 is a sample work request as filled out by a facility user.

There are few variations for this simple step. In very sophisticated systems, the facility user's computer terminals are linked to the maintenance computer, allowing for direct user entry of data.

Transmitting: The work request becomes a work order during the evaluation process described in Chapter 6. Once the work order has been approved for execution, it is transmitted to the shop foreman, or for smaller organizations, directly to the maintenance craftsman. Usually this transmittal is made manually by dispatching the hard copy original work request to the shop or worker. If a computer is utilized, a computer-generated work order is sent to the worker in place of the original work request. Figure 9.2 is a sample computer-generated maintenance work order developed from the handwritten request shown in Figure 9.1.

REQUEST FOR MAINTENANCE SERVICES

Maintenance Control Number|_____

Description of Requested Services

Door closer to storage room is broken again.

Desired Completion Date_____

Work Location		Requested By	Phone	Date Submitted
Bldg.	Room			
Jones	603	I.M. User 253-1234		01/10/88

For Reimbursable Work Only | Authorizing Official
Account to be Charged
_____|_____

BELOW PORTIONS TO BE COMPLETED BY MAINTENANCE CONTROL BRANCH

Priority	Estimated Man-hours	Estimated Mat'l.Cost	Approved By
			Date

Special Instructions

Completion Data |Remarks
 Man-hours Expended   Mat'ls. Used|

Work Completed By	Date Completed	Customer Approval

Figure 9.1

REQUEST FOR MAINTENANCE SERVICES

Maintenance Control Number|_____

Description of Requested Services

   REPAIR BROKEN DOOR CLOSER (STORAGE CLOSET DOOR)

Desired Completion Date_____

Work Location		Requested By	Phone	Date Submitted
Bldg.	Room			
JONES	603	MR. I.M. USER	253-1234	01/10/88

For Reimbursable Work Only	Authorizing Official
Account to be Charged	

BELOW PORTIONS TO BE COMPLETED BY MAINTENANCE CONTROL BRANCH

Priority	Estimated Man-hours	Estimated Mat'l.Cost	Approved By
2	2	10	GHM
			Date
			01/11/88

Special Instructions

   THIRD TIME REPAIR....CLOSER PULLS LOOSE.
   USE HEAVIER BOLTS AND WASHERS.

Completion Data	Remarks
Man-hours Expended Mat'ls. Used	

Work Completed By	Date Completed	Customer Approval

Figure 9.2

The transmittal of work orders can be enhanced by electronic transmission of the data. This is done through remote printers located in each of the shop areas. The work order is dispatched electronically by the trouble desk after approval, estimates, and priority are determined. The exact copy of the work order is printed out in the appropriate location, eliminating the need for manual handling. This feature is particularly helpful where maintenance control and maintenance execution staffs are geographically separated. For emergency work orders, it is necessary, however, to ensure that the work order has been received at the remote site. Oral communication by phone, radio, or beeper is usually desirable.

Updating: Once maintenance work has been performed, the results of that effort are directly recorded on the work order by the maintenance worker. This data is reviewed and verified by the shop foreman and forwarded to maintenance control for updating of the original work order entry. When this data is input into a computer, records are updated and analyzed.

Few computer systems handle this task in any refined manner. It is possible for the maintenance shop to directly update the work order data at a remote terminal. However, only a very large organization will benefit from this feature. Usually the volume of data to be entered can be handled by the work reception/trouble desk clerk during his or her spare time.

Reporting: The results of an individual work order are important to several parties. The shop foreman is concerned with the total listing of incomplete work orders for each particular shop. The facility user is concerned with the status of work requests for their respective spaces or buildings. The maintenance manager is concerned with the timeliness of response to individual work orders and the relative workload of each shop. These multiple demands are handled by repeating the data from the total listing of work orders in different formats, sorted into different categories, and listed in different orders.

The flexibility of any computerized maintenance management system being considered is most evident in this area. A printout of subsets of the total backlog of work orders should be available for the following categories:

- All work orders for each shop
- All work orders for each building or facility area
- All work orders due for completion by a specific time
- All work orders completed by a specific craftsman
- All work orders of a common type — repair, replacement, modification, etc.
- Work orders by differing dollar value and man-hour requirements
- Work orders for common equipment types
- All completed work orders for the above categories
- All pending work orders for the above categories
- All work orders which are overdue for the entire facility
- All work orders which are overdue for a particular shop

It is also desirable to sort each of the above reports in order by some criteria. Common sorting options include:

- By due date
- By shop
- By craftsman
- By priority

- By dollar value
- By date of request
- By work type

In addition to the standard reports listed above, it is desirable to have an *ad hoc query capability* for report generation. This feature, common to many database systems, allows for the generation of most any custom report needed by the maintenance management staff. A sample report, which does not fit the standard categories listed above might be:

A listing of all incomplete repair work orders of priority 2 or higher which are assigned to shops A, B & D.

The ability of computerized systems to handle this type of request varies considerably. In some sophisticated systems, this request may be handled by simple English commands typed in by the maintenance manager. In less sophisticated systems, such a report would require the direct programming of the computer by a trained computer programmer. Since the cost of a computer software system increases dramatically with increases in flexibility, the prospective purchaser should consider the possible report options carefully and determine the minimum requirements for the specific maintenance program.

Planned Work Processing

Planned work occurs at regular intervals and has a predictable cost and duration. Preventive maintenance orders and other recurring jobs are the most common categories of planned work. Once a specific task is recorded in the computer and its frequency established, the computer recalls that data and automatically generates a requirement for its timely completion. Ensuring that this necessary work is not forgotten is the primary asset of a computerized maintenance management system. Planned work must be scheduled, monitored, general progress reports made, and supplies ordered. A computer can greatly enhance these functions, as described in the following section.

Recording: Once identified by the maintenance staff, planned work orders are entered into the computer manually. Figure 9.3 is a typical planned work order, or Preventive Maintenance Order (PMO), as entered by the work receptionist/trouble desk clerk. (Format and content of the PMO is explained in greater detail in Chapter 11.) Various systems provide varying PMO formats, but initial entry of all PMO's and recurring work orders is a one-time, labor-intensive task. The planned work order may be entered by the work reception/trouble desk clerk on a time available basis or, if necessary, contracted to temporary help.

A few computerized maintenance management systems have pre-programmed Preventive Maintenance Orders which can be customized or adopted as is. This feature can save a considerable amount of data entry time. It may also call attention to common PMO's which might not have been identified for the particular facility and equipment being maintained.

Scheduling: Computer scheduling involves the automatic recall of individual work items from a memory. The work items, usually Preventive Maintenance Orders, are printed when they are due for completion. Simple programs search the backlog of PMO's and print those due for work, not taking into account the available staff to complete the work. In such cases, the workload is balanced from one period to the next by input of the due date and subsequent frequency at the time of original data input.

PREVENTIVE MAINTENANCE ORDER

Description

COOLING TOWER MONTHLY SERVICE

Equipment Data

Name	Model No.	Manufacturer
COOLING TOWER	MT-100	HARRIS-BUNSEN

P.M. Priority	Frequency	Due Date
REQUIRED	MONTHLY (DURING COOLING SEASON)	05/01/88

LOCATION

Building	Room	Other
ADMINISTRATIVE	1311	ROOF MOUNT (ACCESS THRU PENTHOUSE)

TOOLS REQUIRED

NORMAL HAND TOOL KIT
CHEMICAL TEST SET
VOLTMETER

MATERIALS REQUIRED

Quantity	Description	Stock Location
1	FAN BELT #1234	A/C SHOP

SAFETY PROCEDURES

SECURE ELECTRICAL POWER AT PANEL AND TAG OUT

MAINTENANCE PROCEDURES

1. CHECK OPERATION OF FEED EQUIPMENT.
2. CHECK BLOW DOWN DEVICES, NOZZLES, FLOAT VALVE.
3. TEST CHEMICAL ADDITIONS. CHECK CONDUCTIVITY.
4. CLEAN SUCTION SCREEN.
5. INSPECT FOR ALGAE GROWTH, DIRT, DETERIORATION OF TOWER.
6. CHECK FAN-ALIGNMENT, TEMPERATURE, LUBRICATION.
7. INSPECT ELECTRICAL CONTACTS FOR PITTING, CORROSION.
8. REPLACE FAN BELT, IF WORN.

COMPLETION DATA

Date Completed	Completed By	Man-hours Expended	Foreman Initials

Craftsman Remarks

Figure 9.3

The output can be made in several forms. The simplest is a listing of PMO's to be done. Armed with the listing, the shop foreman pulls individual PMO instructions from hard copy files and assigns them to a worker. In this system only the title of the PMO, some identifying number, the due date, and frequency are stored in the computer. More elaborate systems will print the entire PMO in a format similar to that shown in Figure 9.3.

The scheduling provisions of a maintenance computer program can provide great assistance to the maintenance manager and shop foremen in balancing the workload for their work force. Since there is a varying level of unplanned work that has to be accommodated, the amount of planned work which should be scheduled each week will vary. The computer program which simply prints the listing of all work needed places the burden for balancing this workload on the foreman. More sophisticated programs will perform this balancing function by analyzing the total workload and printing a schedule (of planned and unplanned work) which fully utilizes the available staff. These programs combine the current unplanned and planned work items, in order of which should be done first. These choices are made by the computer using input management standards, considering the overall maintenance goals and the individually assigned priorities for planned and unplanned work orders.

Monitoring: Once planned work has been assigned, it should be tracked through to completion. Computer records are updated to reflect the time of completion and the man-hours and materials expended. Figure 9.4 is a sample Preventive Maintenance Order completed by the worker and returned for computer updating. Most computer systems have the capability to print a listing of all planned work which was not reported as completed, or deferred maintenance items.

In addition to compiling a record of completed and incomplete work, the computer may be able to update individual equipment histories with the results of each PMO, depending on the sophistication of the computer and the accompanying software. Since additional needed work may be uncovered in the course of completing a PMO, a provision in the computer system might also allow for automatic generation of a new unplanned work order.

Reporting: The planned work reporting function of computerized management systems are similar to those previously described for unplanned work reporting. Listings of planned work by the numerous categories, sorted by the various attributes are desirable. In addition to these reports, a single report of all PMO's for each particular piece of equipment should be provided. As mentioned with unplanned work reporting, the feature of ad hoc query is very desirable for reporting of planned work.

Ordering: Since each planned work activity requires known types and quantities of materials and supplies for accomplishment, the ordering of these supplies can be tracked by the computer, based on the scheduling of the PMO or recurring work activity. The computer may compile a list of required materials for individual or collective work activities for defined periods. Such a listing could be made for all known material and supply requirements for the next three periods. This listing ensures that sufficient materials are ordered well in advance.

PREVENTIVE MAINTENANCE ORDER

Description

COOLING TOWER MONTHLY SERVICE

Equipment Data

Name	Model No.	Manufacturer
COOLING TOWER	MT-100	HARRIS-BUNSEN

P.M. Priority	Frequency	Due Date
REQUIRED	MONTHLY (DURING COOLING SEASON)	05/01/88

LOCATION

Building	Room	Other
ADMINISTRATIVE	1311	ROOF MOUNT (ACCESS THRU PENTHOUSE)

TOOLS REQUIRED

NORMAL HAND TOOL KIT
CHEMICAL TEST SET
VOLTMETER

MATERIALS REQUIRED

Quantity	Description	Stock Location
1	FAN BELT #1234	A/C SHOP

SAFETY PROCEDURES

SECURE ELECTRICAL POWER AT PANEL AND TAG OUT

MAINTENANCE PROCEDURES

1. CHECK OPERATION OF FEED EQUIPMENT.
2. CHECK BLOW DOWN DEVICES, NOZZLES, FLOAT VALVE.
3. TEST CHEMICAL ADDITIONS. CHECK CONDUCTIVITY.
4. CLEAN SUCTION SCREEN.
5. INSPECT FOR ALGAE GROWTH, DIRT, DETERIORATION OF TOWER.
6. CHECK FAN-ALIGNMENT, TEMPERATURE, LUBRICATION.
7. INSPECT ELECTRICAL CONTACTS FOR PITTING, CORROSION.
8. REPLACE FAN BELT, IF WORN.

COMPLETION DATA

Date Completed	Completed By	Man-hours Expended	Foreman Initials
05/10/87	M.J. Pepper	2.0	

Craftsman Remarks

Belt not worn, returned replacement to stock

Figure 9.4

The capability to identify material requirements may be absent from many computerized systems. Sophistication in this area may also include an automatic updating of estimated inventory levels of key supplies, materials, and spare parts. This involves internal computer file linkage between the planned work and inventory control data.

Facility Histories

As individual planned and unplanned work activities are completed, the data regarding the date of completion, man-hours expended, material, supplies, and spare part expenditures are recorded. This data forms a *maintenance history* for the facility, the individual building, and the specific equipment upon which the work was performed. Computer enhancements of these functions are described in the following sections.

Building Data: The long-term observation of maintenance costs by facility management can significantly enhance the forecasting of future maintenance costs. While planned work is definitive in scope and frequency, unplanned work is not; as a facility ages, the cost of maintenance is likely to increase. Facility discrepancies requiring maintenance action may increase in frequency, but in a subtle nature that is not readily observed. Records kept by the computer, tracing individual actions to each building or large component space within a building allows the manager to observe trends indicating such an increase in maintenance costs. The extent to which these trends are traceable depends on the level of definition of the location of the individual work activities. In order to ensure maximum traceability, specific locations of the work must be defined by building, floor, and room.

Since the above-mentioned data is easily captured, most systems have some format for reporting total building costs. Enhancements in this area include automatic calculation of costs per square foot, costs per maintenance visit, cost by various maintenance activity type, etc. This data may also be described graphically, providing visual evidence of any trends.

Equipment Data: Most computer systems are capable of storing the maintenance history of specific pieces of equipment. These equipment histories are valuable tools in predicting times for replacement and for tracing possible maintenance-related causes for equipment failure. With such specific records, the maintenance manager is well informed when deciding whether to repair or replace a failed piece of equipment. The equipment history is a chronological compilation of the maintenance activities of initial installation, preventive maintenance, major and minor repairs, rebuilding, and eventual replacement. Specific parts replaced are cited directly or implied by the notation of a completed Preventive Maintenance Order which contains routine replacement of parts.

The ability of the computer to consolidate and report maintenance activities by specific pieces of equipment is a function of how well individual work activities are recorded. Preventive maintenance order records automatically link work to equipment, but unplanned work must be noted separately on work orders. Enhanced systems link these events together automatically, eliminating the need for dual entry of data.

Maintenance Cost Accounting

Cost is a continual consideration, since manpower utilization, material consumption, and facility down time all require expenditure of company resources or represent lost profits. In order to track specific facility and equipment maintenance costs, records of maintenance activities are kept by individual activity every day. Costs accrue, however, on a less regular basis. Payrolls are made at weekly or bi-weekly intervals. Materials and supplies are purchased in bulk. Spare parts are purchased infrequently.

Due to the difference in the time that resources are expended and when they are paid for, the maintenance management computer system may not be able to make a direct comparison between individual activity cost and actual dollar expenditure. Such a linkage is not mandatory. The system should, however, be able to report costs which *approximately* total the actual expenditure of funds through the payment of payrolls and material invoices. Accurate dollar costs of individual activities are only necessary when such work is to be reimbursed.

To consolidate two separate sets of records (one for maintenance activity and one for maintenance costs) to one cost accounting system is a major evolution with few virtues to justify the tremendous difficulties involved. Exact financial accounting deals with finite instruments such as timecards, purchase orders, and invoices. Control of these items can be left with a few individuals, allowing for greater accuracy of records. If the costs were linked and accounted for exactly at the individual maintenance activity level the number of people involved, and hence the chance for error, increase considerably. For these reasons most computerized maintenance management systems track maintenance costs in parallel with a formal accounting system, but do not usually directly link and reconcile the two systems.

Labor Utilization: Labor is tracked by the number of man-hours expended upon an individual maintenance activity. The total of all man-hours expended on activities during a week should equal the total number of man-hours available and the number of man-hours paid for in payroll. Attempts are made to reconcile any differences. The actual reported hours are retrievable in reports of completed work, by categories of shop, type of work, etc. Computerized systems generally allow the name or code of the worker to be recorded within the computer records for each maintenance activity. This proves helpful when tracing personnel-related maintenance problems.

The degree of accuracy of the total man-hour cost varies with each system. Sophisticated systems require input of actual employee names. The computer then links the number of hours worked to each employee's rate of pay to determine an actual, exact cost for the labor expended. However, this degree of accuracy is not always necessary. These sophisticated systems are most useful when large amounts of maintenance work is conducted on a reimbursable basis.

Equipment and Tool Costs: The cost of using normal shop tools is not tracked to individual maintenance activities. Computer systems tabulate some tool costs in the cost of completing a work order, usually only when tools are specifically rented or purchased for one time use.

Material Costs: Material costs include consumable supplies, raw construction materials, and specific spare parts. Consumable supplies are not usually charged to individual work orders unless used in major quantity or used exclusively for this one work order. If consumables are drawn from stock which was purchased in bulk, the cost can be pro-rated. However, such costs are usually included in overhead costs.

The estimated cost of raw construction materials is charged to specific work orders. Although care is taken to estimate the most accurate costs for raw materials, exact inventory control and costing is not usually worth the time necessary to determine the exact prices and quantities. Spare parts, however, specifically required by a single piece of equipment, are charged to that equipment through exact costing on all work orders. Figure 9.5 is a sample completed work order for work that is not reimbursable, showing material consumption noted by the worker(s) and the shop foreman's estimated cost for those materials entered.

Tracing material costs to a work order can be done with great accuracy. Enhanced computer systems provide direct linkage between individual work orders and the inventory control system to produce absolute material costs.

Inventory Control

The key to prompt maintenance response is the availability of sufficient resources to perform the needed work. Additional personnel can be hired or diverted from other tasks. Tools can be rented. Materials, whether they are consumable supplies or specialized spare parts, however, are not always readily available. For this reason, once the type and quantity of materials to be maintained on hand is defined, a system must be established to determine the availability and location of these materials without extensive search. Most computer systems have provisions for tracking materials through a formal inventory control system. In its simplest form, such capability is limited to an independent listing of item name or part number, and the respective quantity on hand.

Since inventory control is the tracking of the exact number of parts in stock, the sophisticated system will coordinate purchases with increases to the computer inventory levels. As materials are consumed, the reported consumption is noted and reflected in the resulting inventory levels. For rapidly consumed materials, some inventory systems may identify low levels and automatically generate orders for the proper quantity. If desired, the computer may be able to use actual and exact material costs to compute the final reported cost of a work order. For such a system, the worker simply cites the material type and quantity, and the attendant cost is extracted from the computer inventory records.

Any computerized inventory control system is accurate only if regularly verified by comparison of the computer records with actual on-hand stock. The frequency of this comparison is determined by the facilities maintenance manager, based on actual consumption.

Maintenance Management Reports

The maintenance management staff usually desires various reports to measure the overall effectiveness of the maintenance program. These reports will indicate any trends in changing maintenance costs or types of maintenance. A typical report is shown in Figure 9.6. This report shows the status of backlogs for each shop. Useful to maintenance management and shop foremen alike, shop summary reports are consolidations of the total maintenance effort for fixed periods, sorted by individual shops. Figure 9.7 is an example of a shop summary report for a single week. Figure 9.8 is an example listing of shop summary totals for several consecutive weeks.

REQUEST FOR MAINTENANCE SERVICES

Maintenance Control Number|_____
Description of Requested Services

  REPAIR BROKEN DOOR CLOSER (STORAGE CLOSET DOOR)

Desired Completion Date_____

Work Location		Requested By	Phone	Date Submitted
Bldg.	Room			
JONES	603	MR. I.M. USER	253-1234	01/10/88

For Reimbursable Work Only	Authorizing Official
Account to be Charged	

BELOW PORTIONS TO BE COMPLETED BY MAINTENANCE CONTROL BRANCH

Priority	Estimated Man-hours	Estimated Mat'l.Cost	Approved By
2	2	10	GHM
			Date
			01/11/88

Special Instructions

  THIRD TIME REPAIR....CLOSER PULLS LOOSE.
  USE HEAVIER BOLTS AND WASHERS.

Completion Data		Remarks
Man-hours Expended	Mat'ls. Used	*Closer arm bent -*
4.0	$40.00	*Replaced & secured*
		with heavy bolts

Work Completed By	Date Completed	Customer Approval
N.L. Ripper	*01/14/88*	*JMW.*

Figure 9.5

Management reporting capability is of great importance, since all data is gathered with the intent of enhancing management functions and reducing maintenance costs. This area is also subject to varying needs by varying managers with different management styles and intentions. Again, the ad-hoc query feature is most important, allowing for the generation of reports of all types.

General System Features

Previous sections have described the basic structure and functions of a computerized maintenance management system. Both common and desirable features have been outlined. Without attempting to define and explain the nuances of computer programming and database design, there are additional, and more general, features of various programs that merit strong consideration when selecting a system.

Data Integrity

Computers have no inherent intelligence. When a person sees the Carpenter Shop described as "Carp Shop," "Carpentry Shop," or even misspelled as "Carpentre Shop," it is obvious that all of these mean the same thing. The computer, however, recognizes these as *four different shops.* Since data are grouped by many different categories based on entries such as shop name, building name, or work type, all such entries should be consistent. Extra care taken by the data entry clerk can

```
                    SHOP BACKLOG STATUS REPORT
```

| | | | BACKLOG IN SHOP-WEEKS OF WORK: | | |
SHOP NAME	BACKLOG EXPRESSED IN: MAN-HOURS	MATERIALS	CURRENT WEEK	PREVIOUS WEEK	TEN WEEKS PREVIOUS
JANITORIAL	300	2500	2.5	2.7	4.5
PLUMBING	1075	27600	4.7	4.8	4.5
HEATING	675	11300	2.1	2.0	3.4
GROUNDS MAINTENANCE	125	2400	1.0	1.0	.9
AIR CONDITIONING	550	9500	3.5	3.4	5.6
CARPENTRY	2750	18600	7.5	7.8	7.6

Figure 9.6

SUMMARY OF MAN-HOUR UTILIZATION BY SHOP FOR THE WEEK ENDING

6/20/87

SHOP/ MAN-HRS/ PERCENT	HOUSE-KEEPING	GENERAL MAINT.	REPAIR	PREVENT. MAINT.	REPLACE-MENT	MODIFI-CATION	IMPROVE-MENT	SUPER-VISION	INSPEC-TION	ORDERING INVENTORY	TOTAL WORKED	VACATION	GRAND TOTAL
JANITORIAL													
MAN-HOURS	240.0	100.0	30.0	0.0	0.0	0.0	0.0	30.0	20.0	20.0	444.0	0.0	440
PERCENT	54.5	22.7	6.8	0.0	0.0	0.0	0.0	6.8	4.6	4.6	100.0		
PLUMBING													
MAN-HOURS	10.0	30.0	125.0	200.0	18.0	27.0	30.0	40.0	30.0	15.0	525.0	35.0	560
PERCENT	2.0	5.7	23.8	38.1	3.4	5.1	5.7	7.6	5.7	2.9	100.0		
HEATING													
MAN-HOURS	30.0	30.0	250.0	127.0	41.0	10.0	0.0	35.0	9.0	38.0	560.0	40.0	600
PERCENT	5.3	5.3	44.6	22.6	7.3	1.8	0.0	6.3	1.6	6.3	100.0		
GROUNDS MAINTENANCE													
MAN-HOURS	60.0	135.0	15.0	10.0	0.0	0.0	0.0	10.0	0.0	10.0	240.0	0.0	240
PERCENT	25.0	56.3	6.4	4.1	0.0	0.0	0.0	4.1	0.0	4.1	100.0		

Figure 9.7

SUMMARY OF MAN-HOUR UTILIZATION FOR PLUMBING SHOP

EIGHT WEEK PERIOD ENDING 12/04/87

WEEK ENDING	HOUSE-KEEPING	GENERAL MAINT.	REPAIR	PREVENT. MAINT.	REPLACE-MENT	MODIFI-CATION	IMPROVE-MENT	SUPER-VISION	INSPEC-TION	ORDERING INVENTORY	TOTAL WORKED	VACATION	GRAND TOTAL
10/16/87	10	32	150	200	0	24	40	40	28	20	544	16	560
10/23/87	20	18	144	186	32	48	18	40	32	24	522	38	560
10/30/87	12	30	130	192	0	48	20	40	24	18	514	46	560
11/06/87	16	32	115	210	20	24	10	40	32	16	515	45	560
11/13/87	10	30	125	200	18	27	30	40	30	15	525	35	560
11/20/87	12	30	150	180	0	40	10	40	35	17	514	46	560
11/27/87	10	20	90	160	0	25	10	32	26	10	383	177	560
12/04/87	15	35	120	190	25	24	30	40	24	20	523	37	560
TOTAL	105	227	1024	1518	95	260	168	312	231	140	4040	440	4480
AVERAGE	13.1	28.4	128	189.7	10.6	32.5	21	39	28.9	17.5	505	55	560

AVERAGE TIME TO COMPLETE WORK ORDERS:

EMERGENCY - 0.7 DAYS
PRIORITY 1 - 1.1 DAYS
PRIORITY 2 - 3.6 DAYS
PRIORITY 3 - 8.7 DAYS

PREVENTIVE MAINTENANCE WORK COMPLETION RECORD:

ON TIME   - 78%
WITHIN 7 DAYS - 14%
DEFERRED   - 8%

REPAIR TO PREVENTIVE MAINTENANCE RATIO: .677

AVERAGE TIME AVAILABLE EACH WEEK FOR UNPLANNED WORK: 192.1 MAN-HOURS
(TIME SPENT ON REPAIR, REPLACEMENT, MODIFICATION, IMPROVEMENT)

Figure 9.8

211

eliminate many, but not all errors. More sophisticated computer programs have cross-checking features where data are double-checked at the time of entry against specifically allowed data entries. If, for example, the name of a shop does not match one in a previously defined list of shop names, the clerk is alerted that the entry is invalid and must be corrected. Individual entries may also be automatically rejected by the computer if the entry does not match a pattern (such as all numbers or all letters) or if the entry is not within a range of allowed values. This feature reduces the chance that records might be entered or classified incorrectly, misinterpreted, or lost.

Archiving Data

When a computerized maintenance system is purchased, significant investments of time are necessary to enter and update data for the system. Previously kept manual records are disposed of if they duplicate the computer stored records. The computer data are stored on magnetic media, on hard or floppy disks. This media is volatile and subject to unexpected failure, denying access to some or all records. While these failures are rare events, they may be catastrophic unless protection is provided to recover past work. This may be prevented if daily back-ups are made of each file.

Regular Back-ups: Computer hardware and software must provide a system for backing up stored data. This should be done on an external copy to protect the information should a computer failure occur. These back-ups should be made daily, or as often as significant changes are made to the stored data. The entire disk should be backed up once every two weeks. Thus, if a failure occurs the documents can be copied in full (except the day's work) from the back-up tapes or disks. The remainder of the work is then re-entered.

Long-Term Archived Files: As the computer system is used, considerable data accumulates in the permanent memory. As time passes, the need to access this old data is reduced and eventually becomes nonexistent. When such data is no longer used regularly, it should be archived to a floppy disk or tape backup and deleted from the hard disk. This saves room on the disk and speeds the bi-weekly back-up process. Although the need for accessing this old data may seem nonexistent, the cost of the archive disks or tape is very small. If the need does arise, the data can be re-loaded into the computer and accessed very quickly.

System Options

Computerized maintenance management software systems are available in numerous formats which operate on various types of computers. Each system offers different capabilities while following the general topic areas previously outlined in this chapter. The systems fit into three basic categories: off-site systems, package systems, and custom systems. Each of these is described in the following sections.

Off-Site Systems

The off-site system is contracted as a management service. The various characteristics of the facility and its installed equipment are determined by a thorough survey of the facility by the service. The resulting data are entered into a computer located at the site of the software vendor. At regular intervals, the vendor delivers hard copies (printed copies) of preventive maintenance orders to the facility maintenance manager. These work orders are completed by the maintenance staff, reporting the man-hour and material expenditures, and the completed work order is returned to the vendor. The vendor enters the data into the computer and automatically updates the computer maintenance records. Work orders are then usually mailed back to the vendor. These types of maintenance managements services are sometimes called *mail order management services*. This method has the advantage of eliminating the capital investment in computer equipment, and training of computer operators. Sophisticated reports are readily available from these services. The vendors are normally able to tailor the service to exactly match the maintenance procedures employed under a manual record keeping system. The vendors may also offer the services of maintenance professionals who help define necessary preventive maintenance orders and procedures.

Package Systems

Package systems consist of an on-site computer system, complete with software, ready to run. The software has varying capabilities; therefore, the choice of a particular system should be carefully evaluated. When hardware and software are purchased as a package, upkeep of the system can be handled by one firm for both the software and hardware.

There are drawbacks to buying a package system, however. When a system of both software and hardware is purchased, the buyer may lose competitive pricing on the hardware portion of the initial investment. For this reason, hardware and software packages are most often purchased separately.

The quality and capability of these package systems also vary considerably. When a prospective purchaser has existing computer equipment, a demonstration is recommended to evaluate the system's capabilities. It should be noted, however, that when the software is purchased along with the hardware, there is no chance of incompatibility between the computer and the software.

In any case, software packages are written for general application by most maintenance managers, but are not generally customized to the facility's exact requirements, as with off-site and custom systems. However, if the features of a package system match the needs of the maintenance program under consideration, this is usually the least expensive option.

Custom Systems

Custom written systems offer the best quality of service for a computerized maintenance management system. For this type of system, a professional computer analyst designs a set of programs which exactly match the maintenance manager's desired method of record keeping. The manager's desires for prioritized scheduling can be accommodated. Exact reports can be designed and provided. However, the development of *specific* computer programs for a *single* facility application is extremely expensive. And, unless the computer programmer is very familiar with maintenance procedures, the maintenance manager will have to spend significant amounts of time educating the programmer in the company's specific maintenance procedures.

A second option for customized programs is available for persons with some familiarity with computers. Database management software systems are available which can be used to develop a company's own computer maintenance system. However, it is essential that the person developing the system be an experienced computer programmer. Although the currently available database systems are relatively easy to program, the sophistication of an interlinked maintenance management database is not something that should be undertaken by a novice. Custom development with a database system is recommended only for those organizations which (1) possess a competent computer programming staff and (2) are uniquely different in structure or maintenance methods from those supported by package programs.

Hardware Considerations

All computers do not operate with all software; most software is written to run on specific computers or classes of computers. In most cases, an exact match is necessary. Further, maintenance management software is but one of many applications for which a computer is helpful within the daily operation of a maintenance organization.

Existing System

If maintenance management software is to be run on an existing computer system, several factors determine which software can be run on that particular computer hardware. These factors include the type of operating system, the internal memory, and the type of external data storage available.

Disk Operating System: The disk operating system, or DOS, is a standard set of protocols by which a computer accesses programs from a disk. The DOS controls all external devices, sending characters to the video screen or printed output to the printer or disk. Compatible computers employ identical disk operating systems. It is necessary that the software and DOS of an existing computer match. The type of computer on which particular software may run should be explicitly stated in the accompanying literature.

Internal Memory: Computers are manufactured with varying amounts of internal memory capability. For this reason, the memory requirement of a particular software package must match that of an existing computer. In some cases, the computer's memory capacity can be expanded. The memory requirements for each type of software are generally explicitly stated in accompanying literature. Simply having sufficient memory, however, does not guarantee that a program will function properly. An existing program on an existing system may already use portions of the internal memory prohibiting the operation of the new software. It is best to request that the software vendor survey the existing system and make any changes necessary to allow the new software to run properly without diminishing existing computer uses.

External Data Storage: The final factor to consider is the type and amount of *external* memory storage available on an existing system. Most systems have one or more disk drives which accommodate floppy diskettes. These disk drives may not possess sufficient storage capacity to support the large amounts of data inherent in a maintenance management system. Further, these drives are somewhat slow and may make computer operation a tedious process. Usually a hard disk with its greater storage and greater speed for data transfer is required. Most software promotional literature will recommend types and amount of external data storage capability required on the hard disk.

Selecting a System

The maintenance manager should follow the same steps used to solve most engineering problems in the final selection of computer software. The problem must be defined and the parameters of the solution identified. Alternatives are identified and evaluated. The best alternative is selected and implemented.

Selection Parameters

Most firms considering the change from manual record keeping and management to a computer based system will examine several types of software from several vendors. In the course of this search, numerous sales persons are contacted to demonstrate their software. The software they demonstrate may be impressive, but it must satisfy the facility maintenance management needs. The scope of the maintenance organization should be initially stated to prospective software suppliers. They will then demonstrate the manner in which their product is most useful to the particular system.

To present accurate information to prospective vendors, the *size* of the maintenance staff and the facility itself must be defined. This includes providing an organizational chart outlining the number and type of shops and their staffing levels. The type of facility and extent of installed equipment should be summarized.

The *currently employed management methods* for handling work requests, preventive maintenance, cost accounting, inventory control, and management reports should also be outlined.

To provide a complete picture of a company's needs, a complete inventory of any existing computer hardware should be prepared. This should include type of computer, type of disk operating system, internal memory, external data storage devices, and type of printers. Additionally, the current use of the computer should be stated, including the software currently used and whether that software will continue to be used after maintenance management software is purchased and installed.

Establishing Objectives

The maintenance manager should formally describe the expected gains from the implementation of a computerized maintenance management system. These objectives may include goals such as the following:

- reduced downtime due to poor maintenance,
- reduced maintenance costs, and
- reduced permanent maintenance staffing.

Identifying and Evaluating Alternatives

The various types of software available can be identified through search of trade literature or in consultation with various computer suppliers. If there is any doubt as to whether particular software will meet the organizational needs and objectives, sales brochures or demonstration packages should be requested before engaging in a full demonstration.

The maintenance manager should construct a list of required features. This can be done using the features listed in earlier sections of this chapter. Features which are desirable but not required should also be listed. Each possible product should be compared to this list. Figure 9.9 is an example of a checklist of common features and various optional enhancements.

The list should be prioritized to identify those essential, desired features and enhancements which are important for the maintenance program. If a product does not provide these minimum requirements, it should be eliminated from further consideration. Only after the field of choices is narrowed should the cost of the system be considered. The final decision should be cost-based. If the cost of purchasing and converting to a new system is unlikely to produce justifying savings in reduced maintenance, the status quo should be maintained.

**CHECKLIST OF COMPUTER
SYSTEM SOFTWARE FEATURES**

DATE _____

Software–Hardware Considerations

Will software run on existing system _____

Require unique system _____

Support available for unique system _____

Single user system _____

If multi-user, how many users _____

Software support a variety of printers _____

Software Considerations

Unplanned Work

Data Entry

Single point of entry _____

User entry by terminal _____

Data verification available _____

Standard entries sufficient _____

Can entry form be customized _____

Remote printouts supported _____

Convert existing data automatically _____

Maximum number of work orders _____

Estimating

Allow storage of historical cost data for custom use _____

Automatic extentions _____

Multiple wage rates _____

Include overhead _____

Varying overhead rates _____

Link inventory costs _____

Updating Data

Updating directly by shops _____

Customized order for data updating and entry _____

Reports Available

All pending work orders by shop _____

All pending work orders by building _____

All pending work orders by due date _____

All pending work orders by work type _____

All pending work orders by priority _____

All completed work orders by shop _____

All completed work orders by building _____

All completed work orders by due date _____

All completed work orders by work type _____

All completed work orders by priority _____

All overdue work orders by shop _____

All overdue work orders by building _____

All overdue work orders by due date _____

All overdue work orders by work type _____

All overdue work orders by priority _____

Figure 9.9

Multiple sorting options _____

Customized reports _____

Ad-hoc query supported _____

Other features

_____

_____

_____

_____

_____

_____

Planned Work

Data Entry

Standard format for PMO's _____

Standard PMO's available _____

Customized entry forms _____

Maximum number of PMO's _____

Scheduling

Automatic scheduling _____

Consider holidays, non-work days _____

Workload balancing by shop _____

Workload balancing by craftsman _____

Workload balancing by work type _____

Automatically reschedule work _____

Resource allocation _____

Resource leveling _____

Monitoring

Track deferred work _____

Update equipment histories _____

Automatic alert on deferred PMO's _____

Reporting

Same reports as unplanned work _____

Ad-hoc query for PMO's _____

Custom reports for PMO's _____

Material Utilization

Generate long term material requirements _____

Automatic generation of ordering data _____

Coordinate needs with inventory levels _____

Equipment Histories

Contain all parameters:

Type _____

Model number _____

Manufacturer _____

Year installed _____

Track major repairs to equipment _____

Track PMO's to equipment _____

Adaptable to customer needs _____

Figure 9.9 (continued)

Costs should be recorded on cost summary forms such as that shown in Figure 9.10. One form should be completed for each alternative which meets the minimum requirements. It should be noted that computer software is an area where least cost should *not* be the absolute determining factor. Ease of use, availability of training, and degree of after installation support should be given equal consideration.

Implementing the System

Regardless of the advertised advantages of the selected system, little success in the system's use results without firm management commitment to make the system work. That commitment must be a combination of patience, perseverance and understanding. Difficulties should be expected, and system discipline maintained. Worker feedback is also important. These issues are addressed in the following sections.

Expected Difficulties

There are always some difficulties encountered when adopting a new system. The personnel most involved with the new computer system may be frustrated by the interaction with the computer, or not comfortable with the order in which information is requested or printed. Craftsmen may perceive some evil purpose behind the system, such as it being a means of checking their productivity. Certain individuals may be scared that they might somehow damage the computer program and software and refuse to touch the keyboard. A typical pattern of progressive feelings toward the system might be the following:

- stark fear,
- lingering apprehension,
- begrudging acceptance,
- tacit acceptance,
- enthusiastic acceptance,
- desire for system sophistication or improvement, and
- wonderment at how the organization ever managed without the computer.

Each employee who interacts with the computer in any way will welcome the system at one of these progressive feeling levels. It is management's role to ensure that they move *forward* up the chain of acceptance. This is often best handled by example. One enthusiastic individual from any level can inspire similar attempts at acceptance.

In addition to problems with personnel accepting the system, there are often some minor problems with the computer hardware or software. These are normal but frustrating. They can be minimized by requiring full operation and checkout of the system by the supplying vendor(s). Where computers and software are purchased from separate parties, disputes between these suppliers as to probable causes of problems will inevitably arise. These problems can be very damaging to operator acceptance unless they are anticipated as normal events.

Training

Once the commitment is made to purchase the system, the commitment must require its *continued use*, without exception. Training should be offered at *all levels* soon after the implementation. The system operator, usually the work reception/trouble desk clerk, will require the most thorough training. The uses of the various computer generated forms must be explained to shop foreman. The probable use of data collected on completed work orders must be explained to the craftsmen.

MAINTENANCE MANAGEMENT COMPUTER SYSTEMS

COST SUMMARY

ITEM	INITIAL PURCHASE PRICE	INITIAL INSTALLATION COST	ANNUAL MAINTENANCE COSTS
HARDWARE			
COMPUTER - MAIN TERMINAL	-------------	-------------	-------------
ADDITIONAL TERMINALS	-------------	-------------	-------------
HARD DISK DRIVES(S)	-------------	-------------	-------------
FLOPPY DISK DRIVE(S)	-------------	-------------	-------------
TAPE BACKUP SYSTEM	-------------	-------------	-------------
PRINTER #1 _____	-------------	-------------	-------------
PRINTER #2 _____	-------------	-------------	-------------
PRINTER #3 _____	-------------	-------------	-------------
PLOTTER	-------------	-------------	-------------
INTERCONNECTING CABLES	-------------	-------------	-------------
POWER SUPPLY, SPECIAL WIRING	-------------	-------------	-------------
WORK STATION FURNITURE	-------------	-------------	-------------
SOFTWARE			

NOTE: MOST SOFTWARE WILL REQUIRE ANNUAL PAYMENT OF A LICENSE FEE IN ORDER TO RECEIVE ANY UPDATES, CORRECTIONS
      OR CONTINUED TECHNICAL SUPPORT AFTER INITIAL INSTALLATION.

ITEM	INITIAL PURCHASE PRICE	INITIAL INSTALLATION COST	ANNUAL MAINTENANCE COSTS
DISK OPERATING SYSTEM	-------------		-------------
MAINTENANCE MANAGEMENT PROGRAM	-------------		-------------
CUSTOM PROGRAMMING	-------------		-------------
ADDITIONAL PROGRAMS ---			
WORD PROCESSOR	-------------		-------------
SPREADSHEET	-------------		-------------
_____	-------------		-------------
_____	-------------		-------------
_____	-------------		-------------
_____	-------------		-------------
SUPPORT EXPENSES			
PAPER PRODUCTS ---			
STANDARD 8.5 X 11 PAPER			-------------
WIDE CARRIAGE PAPER			-------------
PREPRINTED FORMS			-------------
_____			-------------
_____			-------------
PRINTER RIBBONS			-------------
FLOPPY DISKETTES			-------------
MAGNETIC TAPES (FOR BACKUP SYSTEM)			-------------

OTHER MAN-HOUR EXPENDITURES ----	(ANNUAL)	(RECURRING)	
INITIAL OPERATOR TRAINING	-------------	MAN-HOURS	
INITIAL ENTRY OF EXISTING DATA	-------------	MAN-HOURS	
DATA ENTRY FOR NEW WORK		------------- MAN-HOURS	
DATA UPDATING		------------- MAN-HOURS	
REPORT PROCESSING AND GENERATION		------------- MAN-HOURS	
TAPE BACKUPS		------------- MAN-HOURS	
SOFTWARE UPDATES, SYSTEM TROUBLESHOOTING		------------- MAN-HOURS	
NEW OPERATOR TRAINING		------------- MAN-HOURS	

Figure 9.10

Worker Feedback

Open channels for communication of feedback from employees regarding the computer system must be maintained. Initially, the response to the computer system may be the suggestion that "it will never work." Each must be answered with determined encouragement. As the system gains acceptance, the substance of worker feedback will change to suggested refinements.

- "This report would be more useful if it was sorted by due date."
- "It would be easier to enter the data if the computer requested the building name before asking for the problem description."
- "There isn't enough room on the form to write down all of the materials used on the work order."

These are examples of refinements likely to be suggested. Management should attempt to satisfy these requests. However, the ability to actually implement such changes depends upon the following factors:

- The inherent flexibility of the software to be modified.
- The availability of a programmer or the original software supplier to make the change.
- Any adverse impacts which might accompany the change.
- Management's assessment of the value versus cost of making the change.

Unfortunately, there is no better way to determine the value of a new computer system than accumulated experience. If the decision to select a particular system results from following the procedures outlined in this chapter, problems should be minimal. As experience with a computerized system grows, so will the appreciation of the increased possibilities of more refined or sophisticated systems. A well studied decision will result in satisfaction of the initial objectives which inspired the purchase. If more capability is desired at a later date, the same process of defining desired system capabilities, assessing potential benefits, and estimating probable costs can be employed with continued improvements in maintenance management.

Summary

The selection and purchase of a computerized maintenance management system is a major investment. A need, such as a desire to reduce maintenance costs or improve facility and equipment reliability, is present. Computers offer certain capabilities which can satisfy those needs. Computer programs are readily available and somewhat similar in basic structure. Sufficient differences between programs exist to satisfy a wide variety of maintenance management needs. If management needs are well defined, alternative programs well evaluated, and finally, if the expected benefits exceed the estimated costs, the selected computer system should result in an improved maintenance program.

Chapter Ten

CONTRACTING FOR MAINTENANCE SERVICES

Chapter Ten

CONTRACTING FOR MAINTENANCE SERVICES

For a variety of valid reasons, certain work is best performed by outside companies. Generally, any activity which can be performed by a staff employee can be performed equally well by an employee from a contracted firm. The choice to utilize an outside consultant or servicing firm is, therefore, a cost based decision. The maintenance manager must determine which tasks are best suited for accomplishment by a contractor. Once such a decision is made, the methods for contracting must be determined. This chapter contains descriptions of different methods of contracting. For each one, the required services must be described, the frequency of services established, minimum acceptable levels of quality agreed on, and the schedule of payments defined.

When to Contract for Services

When planning the size and composition of a maintenance work force, the facility maintenance manager's objective is to provide all required services for the lowest cost. These services may be accomplished by employees of the organization or by an independent contractor. If there is sufficient work of a common type to fully occupy a single worker for an entire year, it may be most cost effective to hire a full-time employee. One new employee should be added for each full man-year of common work. The residual work is usually accomplished by private contractors. It should be noted, however, that this analysis assumes that it is cheaper to pay and support a full-time employee than to purchase equivalent services from a private contractor with the same expenses plus an added percentage for profit.

Beyond cost, the need for flexibility and responsiveness often leads to maintaining an in-house maintenance staff, as full-time employees can be re-directed to other work as need arises. In-house employees are also readily available to perform emergency maintenance tasks. If flexibility and responsiveness are not overriding factors, however, and a private company can provide needed maintenance services at acceptable costs, the contractor should be considered.

There are several factors which may allow a contractor to provide needed services at a lower cost than in-house employees, which are listed below:

- Specialization allows higher productivity.
- Specialization allows purchase of expensive servicing or diagnostic equipment.
- Spare parts inventory can cover numerous facilities.
- Familiarity may reduce accidents, lost time, and insurance costs.
- Shop costs are spread over many facilities.
- Employee benefits can be varied with greater flexibility.

The choice between hiring in-house employees and contracting for services is made based on several factors. These are listed below and described in the following sections.

- Frequency of need
- Inadequate in-house talent
- Workload and staff balancing
- Other cost considerations

Frequency of Need

Certain maintenance actions are only necessary on an infrequent or seasonal basis. For example, monthly inspection and preventive maintenance of elevators requires a specially trained craftsman, and few facilities contain sufficient elevators to justify the full-time employment of a qualified serviceman. Similarly, in most regions there is limited need for gardening and lawn services during the winter. Replacement of window glass is another infrequent activity which requires an experienced craftsman to repair. Most facilities have a number of small systems such as these which require specifically trained craftsmen to repair at infrequent intervals. Maintenance costs may be reduced or controlled by identifying these infrequent maintenance activities and considering the cost of contracted services. This should be done during the original structuring of the maintenance work force and during interim reviews of the maintenance effort.

As the maintenance manager carries out major repair projects, such as roof or window replacement, the future quantity of maintenance work may change. Work of a particular type, which was previously sufficient to justify full-time employment of a specific craftsman may consequently be reduced. When a full-time employee of a particular trade specialty is underutilized, the use of contracted services should be considered.

Inadequate In-House Talent

When establishing a maintenance work force and philosophy for a newly constructed facility, the maintenance manager should survey the available work force in the geographic region and develop a plan for in-house and contracted services. If the maintenance manager misreads the labor marketplace and is unable to fill all of the desired positions, some work must be contracted.

Qualified personnel may be obtained by training workers. This alternative, however, should be carefully examined to determine the true cost of providing in-house capability. Aside from the overhead and profit cost additions that accompany contracted services, the raw costs of labor and materials are approximately equal for both in-house and contract performance. The cost of training to provide in-house capability must, therefore, be weighed against the overhead and profit additions associated with the cost of contracted services.

A change to the facility may change the requirements for maintenance talent. For example, conversion from a heating control system consisting of individual room thermostats to modern electronic energy management system may render the in-house personnel unqualified to deal with the new system. In this case, the maintenance manager must weigh the cost of retraining personnel against the cost of contracting outside repair services.

The sophistication of a particular system may dictate that a specialist be hired under contract. Or, state or local law may require a particular level of qualification and licensing for certain work. If in-house personnel do not possess and cannot reasonably obtain such certification, contracted services should be used.

Workload and Staff Balancing

When a particular classification of maintenance work can be performed by in-house personnel, but there is only enough work to justify part-time employment, it may be possible to combine several of these varied work types to justify a single employee known as a *maintenance mechanic*. The combination of work under a single employee's responsibility provides the desired flexibility and responsiveness for unplanned maintenance actions. The total workload will rarely divide into an exact whole number of man-years. Any remainder should be handled by contracted services or overtime rather than adding an underutilized employee to the in-house roles.

Other Cost Considerations

When maintenance employees are added to the organization, they normally accrue all of the *employee benefits* to which the members of that organization are entitled. These benefits include retirement plans, health care insurance, paid vacations, cost of living pay increases, and a myriad of other benefits, each with a specific cost. Even if there is sufficient justification to utilize in-house employees, the cost of providing all of the entitlements may be undesirable and unnecessary. In the unskilled and lesser skilled trades, the cost of contracted services is often significantly less expensive that the total cost of wages plus entitlements for in-house employees. Janitorial and security services typically fall into this category. In many areas there is an available pool of such workers willing to work for wages plus minimal benefits.

Federal, state, and municipal governments have become leaders in the area of contracting for services. Known as privitization, government agencies have established formal programs of turning work over to the private sector through service contracting. Federal government agencies, under the Office of Management and Budget Circular A-76, are required to test the marketplace to see if the private sector can perform any task at a lesser cost than that of performing the work with federal employees. Facility maintenance, janitorial services, and security services are some primary areas in which private sector firms have proven less costly.

In many organizations, an in-place management labor relations agreement may impose restrictive policies and practices which, while valid for the operational side of the organizations, can severely encumber efficient maintenance organizations. These include specific work hours, overtime entitlements, span of control work rules, apprentice programs, work force composition restrictions, and separate lunch rooms. Such policies add to the cost of maintenance, sometimes even requiring extra supervisory positions to oversee them.

The overhead cost necessary to run an in-house organization is the final and major consideration in the decision to contract services. Each group of employees requires supervision, as work must be scheduled and inspected. Work shops should be provided and all materials purchased in bulk stored within the facility. Each employee incurs some payroll costs. Each of these items adds cost to in-house performance.

Types of Service Contracts

Service contracts vary according to duration and the method of payment, based on the ability to predict costs and the known frequency of the required services. The maintenance manager desires that the cost of services be predictable, fixed, and low. The service contractor desires that all costs of providing services are fully compensated and a fair profit earned. The following categories of contracts are grouped by the manner in which the cost of services are established.

Long-Term Fixed Price

Long-term fixed-price contracts are utilized for activities which occur at predictable frequencies and require a predictable quantity of materials and labor. Service contractors can generally accurately estimate the costs for such services and submit a bid. Typically, this type of contract is used for housekeeping, security, trash collection, and routine preventive maintenance. If such services are procured by competitive bidding between selected contractors, the risk to the maintenance manager is minimal and the cost should be reasonable. The contractor, knowing the exact extent of the work, has risks limited to those factors under their control, primarily employee productivity and consumption of supplies.

The long-term fixed-price contract is written for a fixed period, normally one or more years, with a fixed price for that term. For seasonal work, the period may be limited to a lesser duration. The schedule of payments for the services is negotiated between the maintenance manager and the service contractor. Since the term of the contract will run for more than one month, interim payments are usually made.

Long-term fixed-price contracts are not limited to those activities for which exact needs are known. Landscaping and lawn maintenance service contracts may be written generally, such as:

Provide a constant acceptable appearance of the lawns.

In periods of excessive rain, growth may exceed expectations and mowing times are limited; the facility maintenance manager may have to pay for extra mowings, experiencing higher costs than established by the fixed price. When a drought retards normal growth, the maintenance manager is paying for services that are not needed or provided. Such activities are, however, often contracted on a fixed-price basis. Both parties rely upon the predicted average needs to ensure that a fair price is obtained for the maintenance manager and a fair profit earned by the contractor.

The long-term fixed-price contract can also be used for general maintenance activities including minor repairs. An experienced contractor can successfully assess the long-term maintenance requirements for a facility and present a fixed-price bid to provide specified services. Although the exact character of required services may vary, the total quantity of maintenance may be predictable.

While this contracting method is desirable since the costs for the services are fixed for the period of the contract, the maintenance manager does lose some flexibility by committing for a long term. Any desire to reduce the level of services due to a need to redirect funding to more necessary maintenance activities requires modification to the contract, usually at an added cost. The contractor, unless protected by a contract clause, is liable for any increases in the cost of labor or supplies during the term of the contract. Such increases may lead to reduced contractor profits or even losses. Figure 10.1 summarizes the characteristics, advantages and disadvantages of long-term fixed-price contracts.

One-Time Fixed Price

One-time fixed-price contracts are used primarily for maintenance activities which can be predicted well in advance, such as improvement and modification projects. For major repairs where timely response is not required, the fixed price contract is the most economical method. Overhaul of lawn equipment in the winter or snow removal equipment during the summer are examples of repair contracts which can be established at a fixed price. A one-time fixed-price contract in a highly competitive environment provides required services at a known price with minimal risk to the maintenance organization. In such an environment, all risks are assumed by the contractor who is bound to deliver a product or service at the fixed price. The amount of profit which will be earned by a contractor depends on the reliability of the contractor's cost estimate and the degree of competition. Figure 10.2 summarizes the characteristics of the one-time fixed-price contract.

Long-Term Fixed-Price Contracts

Characteristics
Rigidly established scope of work.

Established responsibilities for each party.

A highly specific contract.

Advantages
Firm price eases budgeting.

Single contractor can gain familiarity, improve performance.

Contract is easily inspected and enforced.

Lends itself well to competitive bidding.

Risks assumed by contractor.

Disadvantages
Difficult to change contract.

Changes are costly due to lack of competition.

Difficult to terminate an inferior contractor.

Difficult to fully express needs.

Figure 10.1

Cost Reimbursement Contracts

In order to maintain operation of a facility, a maintenance manager may need specific services immediately. In emergency situations, there is insufficient time to exactly identify the nature of the services and solicit competitive bids for the work. A failed air conditioning system must be repaired, for example, and the cause of the failure and the extent of needed repairs is unknown. In such cases, a contractor is called to diagnose the problem and make the proper repairs.

A contract to reimburse the contractor for all costs, known as a *reimbursement contract*, is used for such emergency repairs. The predicted extent of the repairs and the final cost is unknown. Since making the repair is usually unavoidable, this cost cannot be controlled. The benefit of competition is lost unless labor rates are compared between prospective contractors. Further, the maintenance manager must depend on the contractor to determine the extent of the problem and to make the repairs. Lacking sufficient knowledge to refute or question a diagnosis, the maintenance manager is bound to accept the full cost of the repairs.

These cost reimbursement contracts are often called *time and materials contracts*, since the final cost is simply the total of the labor and material costs. The contractor's overhead and profit margins are either added to the basic labor costs or as a percentage to the raw labor and material costs. Where a maintenance manager has sufficient experience with and trust in a particular service contractor, the cost reimbursement method for services may be used more liberally. Figure 10.3 summarizes cost reimbursement contracts for maintenance services.

One-Time Fixed-Price Contract

Characteristics

A highly specific contract which rigidly establishes the scope of work. Responsibilities for each party are specifically called out. Work is limited to small scope and term and paid for at a fixed price.

Advantages

Allows competitive bidding.

Limited scope allows specialists who should provide lower bids.

Disadvantages

Changes can be costly due to lack of competition.

Difficult to terminate inferior contractor.

Require engineering time and expertise to develop contract documents.

Figure 10.2

Fixed Cost and Retainer

Maintenance of facility equipment and systems usually involves predictable routine preventive maintenance items and minor and major repairs which, while inevitable, are unpredictable. It is highly desirable to have the same contractor perform both the planned and unplanned servicing of these systems. The planned work, being of known quantity, can be performed at a fixed cost. If the unplanned work is performed at a fixed cost, however, the contractor is at risk if needed repairs exceed that fixed cost. The maintenance manager is also at risk of overpaying if actual repair costs are less than the fixed price of a contract. To serve both parties, a contract combining both fixed pricing and cost reimbursement is used. The routine servicing requirements are paid as a fixed cost while repairs are performed at labor rates established in advance.

This method of contracting can provide a desirable level of responsiveness for emergency repairs. This responsiveness is not provided without some cost to the contractor, however. This added cost is paid in the form of a *retainer*. The retainer may be a finite dollar amount or the maintenance manager may agree that all repairs will be performed by the contractor in return for the contractor's guarantee of a rapid response to emergencies.

Cost Reimbursement Contracts

Characteristics

An unpriced contract wherein actual labor and material costs are reimbursed with an additional percentage provided for overhead and profit. Contract can be short or long term. Scope of work not specifically established.

Advantages

Particularly useful for repairing failures of unknown cause.

Can develop scope of work as work proceeds.

Actual costs are fully priceable.

Less risk for maintenance manager.

Contractor can be on the job quicker.

Disadvantages

Contractor has no incentive to provide quick, low cost project.

Requires strict supervision to ensure all costs are "fair."

Lack of competition may yield higher costs.

Figure 10.3

Unit Price

When the type of maintenance activity is well defined but the desired frequency somewhat variable, a *unit price contract* is used. With this type of contract, a price is obtained for one unit of service and payment is made for each unit of service ordered. A unit price lawn mowing contract, for example, would establish the price of a single mowing of all lawns. The maintenance manager would order a mowing only when lawn growth requires it. On a smaller scale, a contract might establish a cost per square yard for steam cleaning carpets. The maintenance manager could than order the cleaning of various offices or rooms without having to estimate the price for each room; payment is simply based upon the quantity of units ordered.

Unit price contracts provide a predictable price for all work and reduce risk for both the contractor and the maintenance manager. Both parties, however, establish upper and lower limits between which the agreed unit price will apply. When the ordered quantities are low, the contractors costs for mobilization and demobilization may raise the unit cost above the agreed unit price. To offset this, contractors often establish a minimum quantity to be provided at the unit price to cover the fixed costs of equipment. In the lawn mowing example cited above, the lawn mowing contractor may be making payments on new equipment. Since the contractor commits to perform all mowing and has to recoup all equipment costs during the fixed mowing season, a minimum number of mowings must be guaranteed and paid for, regardless of the number actually performed. Alternately, when quantities of work are large, the actual unit cost may drop well below the agreed price and the maintenance manager may be paying a higher total cost than is actually necessary. The unit price contract is summarized in Figure 10.4.

A form of unit price contracting is often used to establish a fixed cost for each performance of recurring repair activities. For example, a contract might establish a fixed price for changing one light bulb, another price for changing one ballast in a fluorescent lamp, and a third price for replacing a light switch. Payment is made based on the quantity of each task performed.

Writing Service Contracts

The format for engaging a service contractor is a written agreement between the maintenance manager and the contractor which establishes obligations for each party. The contractor agrees to provide certain services within an established time frame. The maintenance manager agrees to established payments for those services to be made on a specific schedule. The format of the written agreement may vary from a simple statement or even a handshake to a lengthy document spelling out all exact obligations of each party. Some such contracts are written in advance, others on an "as needed" basis. A typical maintenance program utilizes the full spectrum of such contracts.

Establishing workable contracts is a subject worthy of extended study. Years of study and education are necessary to prepare a lawyer to practice contract law. For the purposes of this book, however, the concept and content of a contract can be summarized in a discussion of the basic elements the contract should address. Specifically, the contract should include: a statement of work or services required; the frequency of these services; the level of quality of service required; and the amount and schedule of payments to be made for those services.

Describing Work Requirements

The statement of required work can be made in one of two ways: by specific requirement or by performance requirements.

The exact work may be formally described as a set of specific actions which, when performed by the contractor, will result in the desired level of maintenance. This form of description is called a *specific requirement*. The maintenance manager determines the actions and hires labor to execute those actions. This type of description places the burden of risk on the maintenance manager.

Alternately, the work required may be implied through a statement of the desired end result of maintenance actions. This method is called *performance requirement*. The maintenance manager states the desired performance of the system maintained, and the contractor determines the actions necessary to achieve that desired level of performance. This method places the burden of responsibility on the contractor.

Typical service contracts include some of each method of describing work requirements.

Specific Requirements: A formal statement of exact maintenance actions desired is the best method for stating work requirements. The contractor can readily analyze such requirements, estimate the quantity of labor and supplies necessary to fulfill them, execute the actions, and expect full payment for the work. The responsibility for achieving proper system performance depends on the maintenance manager's ability to predict and describe the required maintenance steps. If the maintenance

---

Unit Price Contracts

Characteristics
> A contract which provides a single fixed cost for each of many types of projects. Payment is made only for actual work performed.

Advantages
> Highly useful for services which will vary in quantity.
>
> Only pay for actual occurrences.
>
> Unit price allows easy ordering of next work.
>
> Easy to track accruing expenditures.

Disadvantages
> Contractor will want guaranteed minimum quantities for each task.
>
> No risk assumed by contractor.
>
> Difficult to define all needed services for unit price bidding.
>
> Scope of each unit price item must be well defined.

Figure 10.4

manager has sufficient knowledge and experience to request the exact amount of work required, the contractor's performance of those actions is simply measured by the contract itself. If the contractor fulfills all stated requirements and the system being maintained fails to perform properly, the maintenance manager is at fault.

Since the risk is assumed by the maintenance manager, great effort is necessary to accurately describe the maintenance actions which result in continued performance of the system. When an activity is being successfully performed by in-house personnel, but is going to move to a contracted method of performance, the development of formal specific requirements may simply be a re-statement of the current methods and procedures. The statement of work required may include all preventive maintenance orders which apply to the maintained system. The same method can be employed for other elements of planned work, where the activities are a known set of steps involving finite labor, materials, and equipment.

Although the specific requirement method lends itself most easily to planned work, certain minor repair work can also be accurately described. The actions necessary to diagnose and cure the cause of a clogged toilet can be listed, knowing the likelihood that such service will be needed. The maintenance history of a facility should be examined to discern those minor repair activities which occur with sufficient frequency to allow for accurate statement of procedures.

Often maintenance service contracts are written for new facilities, equipment, or systems. There are usually no existing procedures or readily available expertise from which to draft the formal maintenance requirements. To begin writing service requirements, the first source for maintenance procedures is the manufacturer's recommended procedures. The formal requirements may also be developed by comparison to other similar systems for which procedures are known. Preventive maintenance procedures for refrigeration equipment, for example, can be modified with reasonable accuracy for a new air conditioning system. The expertise and experience of maintenance craftsmen should also be tapped to produce a reasonable statement of maintenance requirements for new systems.

A newly purchased and installed system often results from a modification or improvement project. When a new system or piece of equipment is purchased or installed, the supplier should be required to provide specific maintenance procedures for that equipment. This is particularly desirable, since a known maintenance cost can be considered and evaluated along with the purchase and installation cost of each considered improvement or modification.

When a major system or piece of equipment fails, the cause and required repair actions are not immediately known. A contractor may be engaged to examine the problem. Once the proposed solution is defined, that contractor may then be directed to make the repairs. The maintenance manager may, alternately, require the initial service contractor to diagnose the problem and provide a statement of exact work required. This work list can then be used to competitively select a contractor to execute the defined repairs. Use of this two-step process is limited to those incidents for which immediate repair is not essential. Additionally, it may not be practical to request a qualified repair contractor to stop work and leave after the diagnosis stage while a search is made for a contractor through competitive bidding.

A final resource for developing a set of specific requirements is the use of a *third party consultant*. A qualified engineer may be engaged to develop a set of maintenance or repair procedures. This method is used extensively for major improvement or modification projects and is described in Chapter 3, "Engineering Considerations for Maintenance."

The specific requirement method for describing work is never a completely thorough statement of required actions. The maintenance manager does not describe the exact physical actions of the servicing craftsman. Manual skills, diagnostic capabilities, and tools required remain unstated and are assumed to be utilized at the level of ability associated with a journeyman level craftsman.

Performance Requirements: It is sometimes desirable to describe only the expected level of performance of a system and to leave the development and execution of specific maintenance actions to the selected servicing contractor. A simple statement of work, such as the following, places the burden of determining the work required (and its attendant labor, material, and equipment costs) upon the servicing contractor.

> Provide all labor, material, and equipment necessary to maintain the installed elevators in acceptable condition and operation.

The maintenance manager is then responsible only for monitoring the operation of the system, rather than for the actual work performance. With this type of description, the maintenance manager is basically unaware of the exact work performed. One drawback of this method is that the maintenance manager relies on a single selected contractor to determine the work and its cost. With this single reliance, the cost may be more than that which would result using the specific requirement format. This extra cost results from lack of competition and from the transfer of risk from the maintenance manager to the service contractor.

When employing the performance requirement method for contracted services, the maintenance manager should obtain, at the commencement of the contract, a statement of the actions the contractor intends to provide for the price stipulated. This allows the maintenance manager to track periodic performance through the term of the contract. For example, in the brief statement concerning elevator maintenance cited above, the contractor might fulfill the requirements by deferring any preventive maintenance tasks and only repairing system failures. For this reason, performance requirements should contain sufficient definition of the desired level of reliability and operation of the maintained system to require a contractor to perform appropriate preventive maintenance.

Actual maintenance requirements may be made implicit in a contract by reference to an *external standard*. For example, where local, state, or federal laws require public inspections of facilities or systems, the contract may stipulate that such work be performed as is necessary to pass that particular inspection. Fire safety inspection by local officials, cleanliness inspections of public eating facilities by sanitation inspectors, periodic elevator and escalator inspections, and manufacturer's warranties are just a few of the possible external standards which may be used to establish a level of performance for a contractor.

The simplest and most frequent use of performance requirement contracts deal with the repair of a system failure. The contract, if any is written at all, calls for restoration of the equipment or system to operating condition. Since it is usually impossible to predict the cost of repairs, cost reimbursement contracts are usually employed.

Responsiveness

After the scope of work is defined, the schedule for work completion should be specified. For planned work, the maintenance manager usually provides the service contractor with sufficient notice, and the work may be scheduled to the convenience of all involved parties. Where an improvement or modification project is contemplated, the time frame for completion may also be identified well in advance of contractor selection. The contractor is bound, by the contract, to commence and complete work within these specified times and dates.

Although unplanned work defies finite scheduling, a priority for accomplishment of this work should be established in advance. Simply requiring that all work be completed immediately, or even "within fifteen minutes of notification," would increase maintenance costs beyond necessary levels. For example, total electrical power failures during a work day must be corrected immediately. Cleaning a clogged toilet, where other toilets are available, is not so urgent. Potential problems should be prioritized according to their impact on facility operation. Varying response times are associated with each priority.

When varying responses are utilized, the contract should establish which party should determine the urgency of the repair response. The response criteria in a contract should be reasonable and in line with the actual minimum needs of continued facility operation.

For predictable equipment or system failures, the maintenance manager should have a contractor in mind for the repair in advance. A list of such problems and potential contractors should be compiled from experience and analysis. Where timely response is necessary, a long-term contract and retainer may be utilized, containing provisions that the contractor will respond to the facility with trained and equipped craftsmen within a specific time period following notification for certain repairs. If the marketplace currently supports an abundance of qualified contractors, it may not be necessary to place one contractor on retainer. The time lost identifying and contracting for a repair may be an acceptable expense when measured against the retainer cost.

Quality

The maintenance manager desires a high level of quality in all maintenance work performance. That level of quality should be expressed in all contract documents. For much maintenance work, the quality of a repair may be dictated by the types of parts or materials to be used. In such cases, a requirement to use only original equipment manufacturer parts should ensure a known quality.

Quality, as it relates to the application of a craftsman's skills, may be difficult to define. Restoration of a failed system to proper operation implies, but does not guarantee, that a quality repair has been made. To ensure quality, the service contractor should be required to guarantee their repairs for a specific period. This warranty shifts the burden for determining quality to the contractor. Asking for a warranty on services may increase the cost of that maintenance action.

With a warranty, the maintenance manager should periodically visit the work site to verify that the contractor's performance meets all contractual requirements. Construction contracts should contain extensive statements regarding both the quality of materials used and the required method of installation. If the engineer has selected and specified the proper materials and methods, and if the maintenance manager regularly inspects the work to ensure that these are employed, a quality product will result.

The final, most effective, and least desirable method of obtaining quality is by withholding payment for services if a quality repair has not been made. This method of quality control may produce immediate, but shoddy, work and may result in a lengthy legal battle. It is generally better to stipulate in the contract specific measurable statements of desired quality.

Payment

The next element included in all maintenance service contracts is the amount and schedule of payments for those services. The general format for payments is determined by the type of contract utilized. The amount of monthly work may not be consistent; however, the contractor generally presents a desired schedule for payment. The maintenance manager must then ascertain whether the percentage of work equals the percentage of payment requested. Most typically, the service contractor is satisfied with a monthly payment schedule which equals one twelfth of the total annual fixed contract price. With several such fixed-price contracts, the contractor can sustain a constant level of staffing and work. The maintenance manager also welcomes a uniform cost for ease of budgeting.

Many contractors offer a discount in price if payment is made quickly after presentation of an invoice. Most contracts allow a thirty-day period between completion of work and payment to the contractor. If this period can be reduced, the contractor may offer a percentage discount. "2/10 Net 30" describes a payment policy whereby a two percent discount is offered if payment is received within 10 days, and no discount if payment is made within 30 days. Just as contracts may offer discounts for prompt payment, there may also be interest charges for slow payments. The maintenance manager must assess the cost of providing fast payment against the savings of the discount. A policy should be developed and followed concerning payment timing.

Types of Services Typically Procured

As previously outlined, an analysis of the total maintenance work effort results in a mix of in-house and contracted services. Figure 10.5 is a summary of direct maintenance work and the types of contracts usually employed for those activities.

Although contracting for services is generally considered for direct maintenance work, the overhead, or indirect portion, of the maintenance effort may also be purchased from outside sources. Figure 10.6 is a summary of indirect work and the corresponding contract types.

Summary

A maintenance manager generally develops a maintenance program which utilizes both in-house and contract resources. The development of individual service contracts requires a description of the work desired, either in specific instructions and procedures or as a statement of the desired end performance of the facility or equipment maintained. The maintenance manager should carefully assess the need for immediate response to maintenance problems. This need should be converted into contractual requirements whenever possible. The cost of the work outlined in the contract is a function of both the work involved, the risk assumed by the contractor, and the type of contract. The primary factor which determines the success or failure of contracted maintenance services is the adequacy of the written contract.

Normal Contract Types for Direct Maintenance Work

Work Type (As defined in Chapter 1)	Preferred Contract	Alternative Contract
Housekeeping	Long term fixed price	—
General maintenance	Long term fixed price	One time fixed price
Preventive maintenance	Long term fixed price	Unit price
Repair	One time fixed price	Cost reimbursement
Replacement	One time fixed price	Unit price
Improvement	One time fixed price	—
Modification	One time fixed price	Cost reimbursement

Figure 10.5

Normal Contract Types for Indirect Maintenance Work

Work Type	Preferred Contract	Alternative Contract
Work identification (usually facility inspections)	One time fixed price	Unit price
Cost estimating (for major projects)	Long term fixed price	—
Purchasing, inventory control	(1)	
Cost accounting (maintain records)	Long term fixed price (2)	—
Scheduling (for workload balancing)	Long term fixed price (2)	
Work tracking & monitoring	(1)	
Facility and equipment histories (maintain records)	Long term fixed price (2)	
Engineering	Retainer	Unit price

Notes
(1) Recommended in-house performance for control.
(2) Usually procured as off site computerized record keeping and processing.

Figure 10.6

Chapter Eleven
PREVENTIVE MAINTENANCE

Chapter Eleven

PREVENTIVE MAINTENANCE

Preventive maintenance makes up a major portion of any maintenance effort. These work items are those maintenance activities which are executed to ensure the continuous operation of a facility, system, or piece of equipment. Preventive maintenance activities facilitate continued operation either through direct completion of work which keeps equipment operating, or through the identification of substandard performance or imminent failure of equipment or systems. A formal program is needed to control preventive maintenance activities. This chapter contains descriptions of preventive maintenance activities and methods of identifying necessary preventive maintenance tasks, and outlines the format of a preventive maintenance order.

Levels of Maintenance

A preventive maintenance program consists of numerous activities, each with different levels of importance. Since preventive maintenance is planned work, the staffing of the organization should anticipate the completion of all preventive maintenance work. Major failures, special events, and other unplanned work occasionally interrupts normal preventive maintenance tasks. When this occurs it is necessary to defer or even skip the performance of some scheduled preventive maintenance activities. In such cases, the least important preventive maintenance activity is deferred. It is necessary, therefore, to rank the importance of preventive maintenance activities in order to ensure that the appropriate tasks are deferred and that essential preventive maintenance work continues.

Critical Preventive Maintenance

Any preventive maintenance action which, if not completed on time, will lead to immediate loss of facility function is a critical preventive maintenance task. For example, failure to lubricate the printing presses of a daily newspaper plant may lead to catastrophic press failure, delays in publication, and loss of revenue. Failure to check fuel levels on emergency generators or heating systems can lead to problems ranging from user inconvenience to loss of facility operation.

Preventive maintenance activities are normally scheduled at the greatest possible intervals. They should be performed no sooner or no later than necessary. If the heating system consumes fuel at a maximum rate of one tank every ten days, it would be wasteful to check fuel levels daily. If, however, the tank is not checked on the ninth day after filling, the possibility of running dry becomes very real. Therefore, on the ninth day, checking the fuel level is critical.

Required Preventive Maintenance

Certain preventive maintenance actions, while inevitable, can be postponed with only minor impact on the facility. These activities must be performed and are scheduled at the necessary frequency. For example, changing the oil in a compressor or engine is required to prevent premature failure. The recommended frequency of such oil changes is established by the manufacturer. If the oil change is not made, excessive contaminants in the lubricating oil could lead to internal wear and failure. If the oil change is delayed only for a short period, however, it is likely that there would be no determinable impact on the service life of the equipment. This is classified as a *required* preventive maintenance task. This means that such tasks can be delayed for a short period, but should not be deferred indefinitely.

Discretionary Preventive Maintenance

The final category of preventive maintenance activities are those work items which can be deferred indefinitely. Of course, such tasks would not be included in the total preventive maintenance program if there was not a positive benefit which exceeded the cost of preventive maintenance performance. However, the predicted benefits of *discretionary* preventive maintenance tasks usually barely exceed the cost of performance. For example, preventive maintenance activities which consist of daily visual inspections of equipment condition can be skipped occasionally with no great impact on the system.

Why Preventive Maintenance?

Preventive maintenance tasks are performed whenever the financial impact of failure to perform the activity exceeds the cost of its performance. The need for each individual preventive maintenance action can be categorized by the manner in which it reduces or avoids costs either in the operation of the facility or elsewhere in the overall maintenance program. These categories are described below.

Maintain Operations

The most obvious justification for preventive maintenance is to maintain facility operations. The continued operation of a facility may be directly tied to a preventive maintenance task when it is performed to maintain manufacturing capability, or, indirectly, where failure to perform a preventive maintenance task may result in degraded systems or equipment. For example, failure of an air conditioner in a computer room may disable operations, while failure of an air conditioning system in an office may not completely impede continued function, but may result in reduced productivity of the workers.

Lengthen Service Life

Many preventive maintenance procedures are designed to ensure that an installed piece of equipment will not fail prematurely. Most equipment is designed for a specific service life. The exact duration of proper operation depends on timely servicing throughout the equipment's use. Such servicing usually consists of several minor activities performed on a regular basis. It also includes some major activities, such as part replacement or overhaul, both of which may provide an extension in service life.

Identify Degradation

Since equipment does not perform at "like new" levels throughout its use, the progressive degradation of system performance should be monitored. Certain preventive maintenance activities should be scheduled to detect these subtle changes and to alert the maintenance work force to major changes in system performance which may indicate imminent failure. Such preventive maintenance tasks may reduce overall costs if the discovery of degraded performance leads to a scheduled replacement at a time convenient to continued facility operations, or if the discovery halts the degradation. In simple terms, a leaky oil line could be repaired before causing problems if regular inspections detected such a leak.

Loss Prevention

Certain installed systems exist to protect either the personnel within the facility or the physical facility itself. For example, a malfunctioning fire alarm system has no effect on facility operation until there is a fire. Preventive maintenance performed on this equipment is justified since losses involving the protected persons or facilities are prevented.

Personnel Safety: Any system or piece of equipment within a facility which is installed to prevent bodily injury or death to the facility users should be regularly maintained. Similarly, any equipment which could cause bodily harm or death during malfunction should be maintained. The benefits related to maintaining protective systems are not always quantifiable, nor is it necessary to attempt such quantification. Protective systems such as exit signs, fire alarms, and emergency lighting are required by applicable building codes or standards. The costs of preventive maintenance for these systems are unavoidable.

Facility Protection: Some systems are installed to protect physical property. Sprinkler systems and burglar alarms reduce risk of property loss and also reduce insurance premiums. If the insurance policies do not require proof of maintenance, the original decision to install the protective system justifies the continuing maintenance. Only when the cost of maintenance increases significantly should the maintenance manager reconsider the need for retaining such a system. In such cases, the repair versus replacement decision should include the consideration of removing or abandoning the system.

Compliance with Standards

Numerous aspects of a facility must comply with various local, state, and federal standards. For example, potable water, produced on site, must meet standards for water quality. Installed sewage treatment facilities must meet effluent standards. In manufacturing plants, certain volatile or poisonous substances must be monitored and kept below certain levels. Boilers must meet emission standards. In each of these cases, the equipment or installed system must be maintained to perform at the levels required. The implications for failure to comply range from a first time warning to a complete shut down of operations. Occupational Safety and Health (OSHA) inspections, for example, may result in fines assessed against the facility for failure to meet OSHA standards.

Working conditions for employees of the organization depend heavily on common systems, such as heating, ventilating, and lighting. Labor agreements may impose significant financial penalties if management fails to maintain proper working conditions cited in the agreement.

Identifying Preventive Maintenance Needs

If the maintenance manager was to thoroughly examine each element of a facility, a preventive maintenance activity could be derived which, if executed, would enhance the performance of that element and, hence, improve the overall operation of the facility. However, not all of these procedures provide benefits which justify the cost of the preventive maintenance activity itself. The field of possible preventive maintenance tasks should be narrowed to that minimum level which maximizes the ratio of benefit to cost. There are several methods by which preventive maintenance activities are identified. A solid, cost effective preventive maintenance program should be developed by considering each of the methods described below.

Impact Analysis

One method for determining which preventive maintenance tasks are essential involves the analysis of the impact of not performing the preventive maintenance task. Simply because a system is critical does not also imply that all maintenance tasks on it prevent failure. Each element, piece of equipment, and system should be examined and classified by its impact on facility operation. Those items which are critical to continued safe operation of the facility receive high priority consideration for maintenance.

This analysis is a two-step process. First, the value of the equipment to continued operation is assessed. Second, the potential value of continual maintenance is determined. The financial impact of under maintained equipment failure is compared with the costs of performing the preventive maintenance.

These comparisons may produce a very small maintenance program which concentrates primarily on mechanical and electrical systems. Using impact analysis, housekeeping and grounds maintenance could be determined undesirable, since the direct financial benefits of these types of work are difficult to determine. Impact analysis, therefore, should be expanded beyond direct financial comparisons to include the impact of the preventive maintenance on satisfaction of organizational goals. These are *soft comparisons*, usually generated from a managerial policy or value judgement.

Impact analysis, as a method for identifying preventive maintenance activities, is summarized in the following steps:

1. Consider management goals for facility operation, develop a set of maintenance goals. (This process was outlined in Chapter 1.)
2. Examine the facility in total and identify those elements whose proper performance or appearance are necessary to meet the established organizational and maintenance objectives.
3. Determine the impact (upon those goals) of the failure of each of these component parts of the facility.
4. Determine what preventive maintenance actions might mitigate such potential failures.
5. If the cost of such maintenance actions is less than the impact of failure, place the appropriate preventive maintenance activity on the maintenance schedule.

Failure Analysis

Preventive maintenance activities may be identified by examining the maintenance history of a facility. In particular, each failure of a facility system or piece of equipment offers an opportunity to determine whether preventive maintenance could have averted or delayed the failure. The process of failure analysis is shown in Figure 11.1.

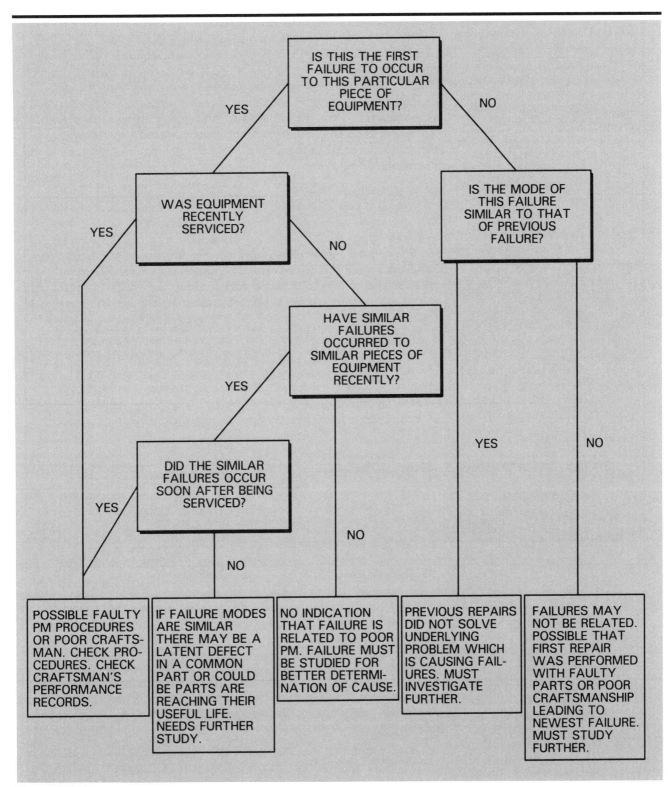

Figure 11.1

Manufacturer's Recommendations

A primary source for identification of preventive maintenance activities is procedures recommended by the equipment manufacturers. Where entire systems or major system components are mass produced, the manufacturer usually recommends service frequencies and procedures. Any warranties provided with such equipment are usually dependent on completion of any recommended preventive maintenance. While such warranties are in force, these tasks are, in fact, required. After warranties expire, the maintenance manager may reconsider the desirability of continuing these activities at the recommended frequency.

Analogous Equipment

When the components of mechanical and electrical equipment are examined, many similarities are revealed. Preventive maintenance procedures which are developed for one electric motor may be readily adaptable for most other electric motors. An electrical switch is maintained in a particular manner, regardless of the equipment which it powers. Such analogies between equipment components allow for easy development of preventive maintenance procedures when the manufacturer fails to provide any instructions.

Analogous development of preventive maintenance is not limited to individual component parts. The principle can be applied for entire systems. This may be helpful in predicting maintenance requirements for a new system being considered for installation. From the costs of preventive maintenance procedures for existing similar equipment, a reasonable estimate of the life cycle maintenance costs of the new system can be developed.

The Preventive Maintenance Order

Once the need for preventive maintenance for a system or piece of equipment has been determined, the actual procedures to be used in performance of that preventive maintenance task should be established. The written directive given to the craftsman performing the maintenance is the *preventive maintenance order* (PMO).

Although the preventive maintenance order describes the actual physical actions of executing the maintenance, it is more than a "how to" guide. The PMO is a planning document, a safety document, a road map, an assistance in ordering parts, and an inspection and feedback tool. Each of these items is addressed in the following sections, in the order that the PMO is developed and used. Figure 11.2 is a sample preventive maintenance order form.

Description

The PMO should include a brief description of the work expected. This is used as a title, particularly important in computerized maintenance systems where it is referred to in reports and scheduling work. Even if the preventive maintenance system is not computerized, the short title should be used for the assignment of work.

Computer systems are designed to accept a limited number of characters in the title. This limit may sometimes challenge the maintenance manager to come up with a descriptive title which captures the essence of the work. The title may describe the specific work required, such as "Adjust compressor belt tensions," or it may describe a number activities under an umbrella title such as "Spring servicing A/C unit" or "Test fire alarm system." The title, or description, will be used by various personnel, therefore, it should be easily understood by all levels, from maintenance manager to the craftsman. Technical terms, known only to the servicing craftsman or technician, should be avoided.

PREVENTIVE MAINTENANCE ORDER

Description
_____

Equipment Data

Name	Model No.	Manufacturer
P.M. Priority	Frequency	Due Date

LOCATION

Building	Room	Other

TOOLS REQUIRED
_____

_____

MATERIALS REQUIRED

Quantity	Description	Stock Location

_____

SAFETY PROCEDURES
_____

_____

MAINTENANCE PROCEDURES
_____

_____

COMPLETION DATA

Date Completed	Completed By	Man-hours Expended	Foreman Initials

Craftsman Remarks
_____

_____

Figure 11.2

Equipment Type

The remaining information on the form should be technically and specifically accurate. The type of equipment or component upon which the preventive maintenance task is performed must be thoroughly described, including the equipment name, model number, and manufacturer. The craftsman will need this information if service manuals or schematics are to be consulted. If the PMO identifies a need for replacement parts, this equipment data will save time in ordering. Where it may have some importance, a serial number may also be recorded. Figure 11.3 shows a PMO with the title and equipment data completed.

Not all PMO's deal with a specific piece of equipment. For example, a PMO for touch up painting a wall, or cleaning a carpet, should note the type of wall surface and substrate, or the carpet material type and pile height, respectively.

Preventive Maintenance Type

The classification of the particular preventive maintenance task should be shown, using levels such as *critical*, *required*, and *discretionary*. This classification should convey information concerning the urgency of the preventive maintenance task. The maintenance manager and shop foreman will use this data when the maintenance work force is temporarily directed away from preventive maintenance performance. Those preventive maintenance tasks classified as least important will be the first to be deferred.

Frequency

Many preventive maintenance orders are repeated weekly, monthly, annually, or at some other interval. This frequency should be noted on the PMO. The craftsman should be aware of the length of time between servicing in order to categorize changes that have occurred to the equipment. The frequency characteristic is used by computerized maintenance system internal programs to determine the next scheduled due date.

Due Date

The due date usually represents the date upon which the work should be performed. As such, it allows little flexibility in scheduling the maintenance work force; however, the use of this date varies. Some organizations regard the date as an exact mandated performance date. Others use the date as the last date by which the work may be performed, allowing shop foreman flexibility in balancing the work load for their craftsmen. Still other organizations use the due date as a signal of when the PMO should be issued to the shop. Thereafter, the shop foreman schedules the work for the immediate future.

Location

The location data for a PMO should contain sufficient detail to lead the craftsman to the exact location of the equipment or system to be serviced. The location for equipment in a facility consisting of a single building may include only floor and room number. Larger facilities with several buildings add the building name or number. Still larger facilities may include the geographic site if the maintenance organization serves several separate sites.

Once within a room, the physical location of the work is usually easy to find. For large rooms, or equipment hidden behind walls or above ceilings, exact physical location should be described.

PREVENTIVE MAINTENANCE ORDER

Description

COOLING TOWER MONTHLY SERVICE

Equipment Data

Name	Model No.	Manufacturer
COOLING TOWER	MT-100	HARRIS-BUNSEN

P.M. Priority	Frequency	Due Date

LOCATION

Building	Room	Other

TOOLS REQUIRED

MATERIALS REQUIRED

Quantity	Description	Stock Location

SAFETY PROCEDURES

MAINTENANCE PROCEDURES

COMPLETION DATA

Date Completed	Completed By	Man-hours Expended	Foreman Initials

Craftsman Remarks

Figure 11.3

Computer systems allow for extensive analysis of the maintenance effort by sorting and calculating data for several locations. If work is charged back to the different divisions of the organization, it is necessary to add that divisional designation to the PMO. Figure 11.4 shows a PMO with type, frequency, due date, and location data added.

Tools Required

When the PMO is delivered to the craftsman for completion, the craftsman assembles the necessary tools for the work. Even after several repetitions of the same PMO, the worker may occasionally forget a needed tool. The PMO is the proper place to provide a reminder. It is especially helpful to note any equipment, such as ladders or keys, which are needed for access to the work location.

Materials Required

The type and quantity of materials, spare parts, or consumables needed for the work should be listed on the PMO. Time lost returning to the shop or warehouse for materials or tools can make a significant dent in productive output. This listing of required materials is also useful to the shop foreman for purchasing and inventory purposes.

Safety Procedures

An active safety program should be part of any efficient maintenance program. Continual classes may be taught, and/or posters placed to warn of dangers or encourage safe practices. Craftsmen in entry level training should be made aware of safe working practices. Since a PMO is written once and used many times, time should be taken to detail specific safety procedures. Corrective or constructive work orders, however, have limited repeatability and the time required to produce exact and individual safety procedures is often not worth the effort.

Safety procedures, as listed in a PMO, should include directions to turn off and lock out any electrical circuits, requirements for safety glasses or hearing protection, or other measures to protect the worker from bodily injury. Additionally, the PMO safety procedures should be used to remind workers to consider the safety of others. Installation of special ventilation, disabling circuits that might be affected by the work, and cleaning the work area when done are examples of such safety guidelines.

User Clearance

Much preventive maintenance work is performed during working hours in the operational spaces of the facility. Since maintenance work is often obtrusive, noisy, and dirty, it may be necessary to obtain clearance from the facility users within the space before commencing work. When electrical power must be secured which will disrupt or stop facility use, the preventive maintenance activity should be scheduled in advance by the shop foreman or maintenance manager. For minor disruptions, the craftsman will be responsible for making proper notice or obtaining permission from facility users as cited on the PMO. An unannounced securing of electrical power can cause significant loss of computer data requiring immense work to reconstruct the lost efforts. Workers are justifiably upset when such power failures occur accidentally, but they are furious when the securing of power is a conscious, unannounced act. Tests of fire alarm systems should also be announced in advance to prevent unintended evacuation.

Estimated Time

If accurate estimates are available to predict the average time required to complete a PMO, this should be included. Such data is useful to the shop foreman for planning a full work day for the employees.

PREVENTIVE MAINTENANCE ORDER

Description

  COOLING TOWER MONTHLY SERVICE

Equipment Data

Name	Model No.	Manufacturer
COOLING TOWER	MT-100	HARRIS-BUNSEN

P.M. Priority	Frequency	Due Date
REQUIRED	MONTHLY (DURING COOLING SEASON)	05/01/88

LOCATION

Building	Room	Other
ADMINISTRATIVE	1311	ROOF MOUNT (ACCESS THRU PENTHOUSE)

TOOLS REQUIRED

MATERIALS REQUIRED

Quantity	Description	Stock Location

SAFETY PROCEDURES

MAINTENANCE PROCEDURES

COMPLETION DATA

Date Completed	Completed By	Man-hours Expended	Foreman Initials

Craftsman Remarks

Figure 11.4

The craftsman can use the time as a guide to judge performance. However, whether to include the time estimate on the PMO given to the worker should be carefully considered by the maintenance manager. The maintenance manager should determine if there may be any ramifications to providing the data. For example, if the estimates are conservative, reflecting worst case situations, the workers may be tempted to slow the work rate to match the estimate. If the estimate is an optimistic duration, obtainable only if everything goes perfectly, the workers would, at first, attempt completion in the estimated time. Continual publication of overly optimistic estimates, however, may cause the workers to simply disregard the numbers.

Maintenance Procedures

The final portion of the PMO is the meat of the document, its reason for being. The statement of actual maintenance procedures. This statement should include sufficient detail to instruct the craftsman to complete all work desired. Since the PMO is given to a skilled craftsman, the instructions should address general, rather than specific, manual tasks. A statement such as, "adjust drive belts to proper tension" is generally sufficient. Thorough step by step instructions of which tools to use and how to use them are not necessary. Specific instructions are needed, however, when the worker may not possess sufficient talent or knowledge to determine the proper course of action or desired end result. In such cases, a description such as "torque all bolts to 100 foot-pounds" is better than "re-torque all bolts."

There are several sources to aid in the development of specific maintenance procedures. The recommendations of the equipment manufacturer may be used as written, or amended for the proper level of detail. The experience of the shop foreman and maintenance manager should be combined to write a reasonable statement of procedures for common maintenance tasks. For existing preventive maintenance tasks being performed by a craftsman without specific guidance, the craftsman may be called on to write the formal procedures. Consultants may be hired to write other specific maintenance procedures.

All written procedures should be periodically reviewed for accuracy. This is important when a piece of equipment fails, despite having been maintained exactly according to the written procedures. When the PMO was written, and then executed several times, the normal assumption would be that the maintenance was sufficient and not the cause of the failure. However, a missing step in the PMO, whether due to ignorance or oversight, could lead a piece of equipment to premature failure.

Figure 11.5 shows the complete printed preventive maintenance order ready for delivery to the shop foreman assignment to a craftsman for completion.

Completion Data

The final section of the PMO is reserved for the performing craftsman to report the results of the preventive maintenance task. In addition to signifying completion of the task, this block can be used to record any problems encountered, any abnormal conditions noted, any materials consumed, and any recommendations for further action. The time required to complete the preventive maintenance activity and the name or signature of the completing craftsman should be recorded.

PREVENTIVE MAINTENANCE ORDER

Description

COOLING TOWER MONTHLY SERVICE

Equipment Data

Name	Model No.	Manufacturer
COOLING TOWER	MT-100	HARRIS-BUNSEN

P.M. Priority	Frequency	Due Date
REQUIRED	MONTHLY (DURING COOLING SEASON)	05/01/88

LOCATION

Building	Room	Other
ADMINISTRATIVE	1311	ROOF MOUNT (ACCESS THRU PENTHOUSE)

TOOLS REQUIRED

NORMAL HAND TOOL KIT
CHEMICAL TEST SET
VOLTMETER

MATERIALS REQUIRED

Quantity	Description	Stock Location
1	FAN BELT #1234	A/C SHOP

SAFETY PROCEDURES

SECURE ELECTRICAL POWER AT PANEL AND TAG OUT

MAINTENANCE PROCEDURES

1. CHECK OPERATION OF FEED EQUIPMENT.
2. CHECK BLOW DOWN DEVICES, NOZZLES, FLOAT VALVE.
3. TEST CHEMICAL ADDITIONS. CHECK CONDUCTIVITY.
4. CLEAN SUCTION SCREEN.
5. INSPECT FOR ALGAE GROWTH, DIRT, DETERIORATION OF TOWER.
6. CHECK FAN-ALIGNMENT, TEMPERATURE, LUBRICATION.
7. INSPECT ELECTRICAL CONTACTS FOR PITTING, CORROSION.
8. REPLACE FAN BELT, IF WORN.

COMPLETION DATA

Date Completed	Completed By	Man-hours Expended	Foreman Initials

Craftsman Remarks

Figure 11.5

This data should be reviewed by the shop foreman who may, in turn, initiate further investigations, make recommendation for further actions, or issue a work order to correct discrepancies, as appropriate. To ensure that the comments have been noted, the shop foreman should initial the PMO. If the craftsman notes changes which should be made to any part of the PMO, such recommendations should be provided to the maintenance control branch, which will evaluate and make corrections to the PMO for future execution. Figure 11.6 is a sample completed PMO with a craftsman's comments.

The PMO Cycle

The PMO is a document with many purposes. It forms the one constant element of a preventive maintenance activity. Figure 11.7 shows, in flow chart form, the cycle of a preventive maintenance order.

Fine Tuning the Program

A preventive maintenance program should be reviewed periodically to determine its effectiveness. That effectiveness should be measured by the performance of the facility and equipment maintained. It is also measured by the relative cost of executing the entire program. In an effort to ensure that no failures occur, a degree of over maintenance can creep into any program. Particular attention should be paid to discretionary and inspection related preventive maintenance activities. If discretionary PMO's are continually deferred or ignored with no apparent effect upon facility and equipment performance, they should be dropped. If inspections rarely note discrepancies, or if the inspections could be performed by the facility user, consideration should be given to deleting or revising the frequency of these inspections. A PMO which requires an electrician to inspect every space on a weekly basis to determine if any light bulbs have burned out is probably a waste of effort, since the users will normally report any extinguished lights. That inspection PMO might be reduced to include only inspection of exit lighting, which is not normally noticed by the user.

In addition to examining the preventive maintenance program for wasted efforts or over-maintenance, the failure record for equipment should be examined periodically to identify possible "holes" in the preventive maintenance program. Major failures may occur at distant intervals and a common tie of two or more failures to inadequate maintenance may be missed. Unplanned maintenance work can be summarized by type of equipment and by the location of work. Trends may be identified which indicate deficiencies in preventive maintenance. Major failures can be thoroughly studied, including determining if a particular craftsman performed all preventive maintenance work preceding the failure. The timing of a failure with respect to the most recent preventive maintenance may reveal that the interval between preventive maintenance is too great or that the failure was caused by an improperly performed PMO.

The character of a failure should be examined to determine if any preventive maintenance could have prevented the failure. If no specific preventive maintenance, including inspection, could have prevented a failure or provided advance warning, the existing PMO's for that equipment should be studied. It is possible that the type of failure which these PMO's are designed to prevent would never occur if the PMO was never executed, since the equipment invariably fails in other ways due to unforeseen causes.

PREVENTIVE MAINTENANCE ORDER

Description

COOLING TOWER MONTHLY SERVICE

Equipment Data

Name	Model No.	Manufacturer
COOLING TOWER	MT-100	HARRIS-BUNSEN

P.M. Priority	Frequency	Due Date
REQUIRED	MONTHLY (DURING COOLING SEASON)	05/01/88

LOCATION

Building	Room	Other
ADMINISTRATIVE	1311	ROOF MOUNT (ACCESS THRU PENTHOUSE)

TOOLS REQUIRED

NORMAL HAND TOOL KIT
CHEMICAL TEST SET
VOLTMETER

MATERIALS REQUIRED

Quantity	Description	Stock Location
1	FAN BELT #1234	A/C SHOP

SAFETY PROCEDURES

SECURE ELECTRICAL POWER AT PANEL AND TAG OUT

MAINTENANCE PROCEDURES

1. CHECK OPERATION OF FEED EQUIPMENT.
2. CHECK BLOW DOWN DEVICES, NOZZLES, FLOAT VALVE.
3. TEST CHEMICAL ADDITIONS. CHECK CONDUCTIVITY.
4. CLEAN SUCTION SCREEN.
5. INSPECT FOR ALGAE GROWTH, DIRT, DETERIORATION OF TOWER.
6. CHECK FAN-ALIGNMENT, TEMPERATURE, LUBRICATION.
7. INSPECT ELECTRICAL CONTACTS FOR PITTING, CORROSION.
8. REPLACE FAN BELT, IF WORN.

COMPLETION DATA

Date Completed	Completed By	Man-hours Expended	Foreman Initials
05/10/87	M.L. Pepper	2.0	

Craftsman Remarks

Belt not worn, returned replacement to stock

Figure 11.6

The Preventive Maintenance Cycle

STEP ONE
Preventive maintenance order (PMO)
written after initial equipment installation
or in response to perceived need for
improved maintenance.

STEP TWO
PMO entered into computer or placed on
shop foreman's schedule.

STEP THREE
Date for next performance of the PMO
approaches. Computer prints notice or need
to perform PMO (or shop foreman notes it
is due if no computer is used).

STEP FOUR
Foreman assigns work to craftsman who
proceeds to complete work. Upon
completion of PMO, the craftsman fills in
completion data on PMO and returns PMO
to foreman.

STEP FIVE
Foreman reviews completed PMO.
Notes and schedules any additional
work recommended by craftsman, records
completion cost data, and returns to
work receptionist.

STEP SIX
Work receptionist enters completion
date and data into computer, which
automatically records completion in
equipment history file.

STEP SEVEN
Computer automatically prints report of
overdue PMO's. Foreman and maintenance
manager determine if PMO can be deferred
and take proper action.

Cycle continues at step three

Figure 11.7

Summary

A sound preventive maintenance program is essential to preclude more expensive failure of equipment, installed systems, or other facility components. The elements of the facility which can benefit from periodic preventive maintenance should be identified. The cost of such maintenance is weighed against the impact costs of not performing the maintenance. For the preventive maintenance activities which are determined to be worth performing, a preventive maintenance order should be written. Additionally, the preventive maintenance program and the records of unplanned maintenance work should be periodically reviewed to reveal needed modifications to the preventive maintenance program.

INDEX

INDEX

NOTES

NOTES

NOTES

NOTES

NOTES

NOTES

NOTES

NOTES

NOTES

NOTES

NOTES

NOTES

NOTES